T0215055

What Every Engineer Should Know About Risk Engineering and Management

Completely updated, this new edition uniquely explains how to assess and handle technical risk, schedule risk, and cost risk efficiently and effectively for complex systems that include artificial intelligence (AI), machine learning, and deep learning. It enables engineering professionals to anticipate failures and highlight opportunities to turn failure into success through the systematic application of Risk Engineering. *What Every Engineer Should Know About Risk Engineering and Management*, Second Edition, discusses Risk Engineering and how to deal with System Complexity and Engineering Dynamics, as it highlights how AI can present new and unique ways that failures can take place. The new edition extends the term "Risk Engineering" introduced by the first edition to Complex Systems in the new edition. This book also relates Decision Tree which was explored in the first edition to Fault Diagnosis in the new edition and introduces new chapters on System Complexity, AI, and Causal Risk Assessment along with other chapter updates to make the book current.

Features

- Discusses Risk Engineering and how to deal with System Complexity and Engineering Dynamics.
- Highlights how AI can present new and unique ways of failure that need to be addressed.
- Extends the term "Risk Engineering" introduced by the first edition to Complex Systems in this new edition.
- Relates Decision Tree which was explored in the first edition to Fault Diagnosis in the new edition.
- Includes new chapters on System Complexity, AI, and Causal Risk Assessment along with other chapters being updated to make this book more current.

The audience is the beginner with no background in Risk Engineering and can be used by new practitioners, undergraduates, and first-year graduate students.

What Every Engineer Should Know

What every engineer should know amounts to a bewildering array of knowledge, as engineering intersects with all fields of modern enterprise. This series of concise, easy-to-understand volumes provides engineers with a compact set of primers on practical subjects such as patents, contracts, software, business communication, management science, and risk analysis, as well as more specific topics like embedded systems design. The books in this series address important problems encountered in the line of daily practice—problems concerning new technology, business, law, and related technical fields

Series Editors
Phillip A. Laplante
The Pennsylvania State University, Malvern, USA

Proposals for the series should be sent directly to one of the series editors above, or submitted to:

Taylor and Francis Group
3 Park Square, Milton Park
Abingdon, OX14 4RN, UK

What Every Engineer Should Know About Business Communication
John X. Wang

What Every Engineer Should Know About Career Management
Mike Ficco

What Every Engineer Should Know About Starting a High-Tech Business Venture
Eric Koester

What Every Engineer Should Know about MATLAB® and Simulink®
Adrian B. Biran

Green Entrepreneur Handbook: The Guide to Building and Growing a Green and Clean Business
Eric Koester

Technical Writing: A Practical Guide for Engineers and Scientists
Phillip A. Laplante

What Every Engineer Should Know About Cyber Security and Digital Forensics
Joanna F. DeFranco

What Every Engineer Should Know About Modeling and Simulation
Raymond J. Madachy, Daniel Houston

What Every Engineer Should Know About Excel, Second Edition
J. P. Holman, Blake K. Holman

Technical Writing: A Practical Guide for Engineers, Scientists, and Nontechnical Professionals, Second Edition
Phillip A. Laplante

For more information about this series, please visit: https://www.crcpress.com/What-Every-Engineer-Should-Know/book-series/CRCWEESK

What Every Engineer Should Know About Risk Engineering and Management

Second Edition

John X. Wang

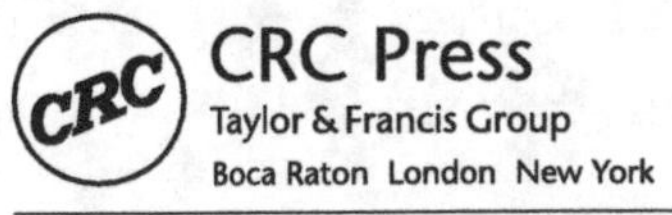

CRC Press is an imprint of the
Taylor & Francis Group, an **informa** business

Second edition published 2023
by CRC Press
6000 Broken Sound Parkway NW, Suite 300, Boca Raton, FL 33487-2742

and by CRC Press
4 Park Square, Milton Park, Abingdon, Oxon, OX14 4RN

CRC Press is an imprint of Taylor & Francis Group, LLC

First edition published by CRC Press 2000

ISBN: 978-1-032-44210-5 (hbk)
ISBN: 978-1-032-43982-2 (pbk)
ISBN: 978-1-003-37101-4 (ebk)

DOI: 10.1201/9781003371014

Typeset in Times
by KnowledgeWorks Global Ltd.

This book is dedicated to:

My mom. Thank you for encouraging me write books to develop my own voice.

My dad. The fact that my dad is still writing research papers on Minkowski geometry at the age of 92 inspires me every time.

Contents

Figures

Tables

Foreword

In aerospace and engineering in general, the major metrics are capability, cost, and safety/risk. With the increasingly rapid emergence and utilization of new technologies and systems of increasing complexity, ensuring safety becomes more difficult. The technology and practice/applications of safety/risk technologies require updating in concert with capability and systems technology changes. Hence the new, updated edition of this risk engineering book. Major changes in technology and applications, now and going forward, increasingly involve artificial intelligence (AI)/autonomy and complex systems, which are rapidly developing and moving targets when it comes to risk/safety analysis. These introduce both new risks, safety issues, and alter more usual ones. The current reality is that often the best AI is when it is "Black Boxed", developed "independently", without detailed human understanding of how and by what processes decisions and results are produced. There are ongoing efforts to make the AI processes more understandable by humans, with results TBD. Also, trusted and true autonomy is free of human intervention, which would require machines to ideate to solve in real time issues that arise due to unknown unknowns and even known unknowns. Machines using generative adversarial networks (GANs) and other approaches are beginning to ideate. Overall, the capabilities and practice of AI/autonomy utilization is a work in progress with the rate of progress substantial and the impacts upon system risk/safety major.

As systems become more complex and have increasing numbers of piece parts, there are possibilities for cascading failures, where subcritical issues with one part can alter the functionality of other system piece parts and cause subcritical failures, contingent upon which/what/how the system is configured. Hence the need for serious risk/safety analyses at the system and system of systems level. In many systems/applications, software is both the most expensive and the most troublesome because of extant errors when installed and errors instilled during operation either by operators or by environmental effects. In many systems, human factors are major sources of error, including traffic accidents and in aerospace. Safety or risk issues in aerospace include rockets, whose safety record is an issue occurring every 100 launches or so, orders of magnitude more of a safety concern than the superb safety record of commercial aviation. There is now interest in certification by computation vice experiment. Unfortunately, the computations do not consider, nor have available, the detailed aspects of the as-built physical system.

Overall, safety/risk is not what it was in the Industrial Age, and as this book espouses, risk considerations now include a vast number of possible issues, with an increasing number of newer technologies and complexities and assumptions, all of which can interact at the systems of system level. All of this is in the context of the increasing importance of cost, which puts a premium upon reducing "fail safe" redundancies and over design as a whole. Risk assessment is both changing rapidly and becoming ever more difficult and critical to engineering success. Hence this updated edition.

Dennis M. Bushnell
HFAIAA, FASME, FRAeS

Preface

Engineering of systems always involves a variety of elements of risk, especially when an undertaking is the first of its kind. A large-scale engineering design can fail in many ways and the failure modes are sometimes difficult to completely anticipate due to the system complexity. Since the objective of engineering is to design and build things right the first time, knowledge of how to manage technical risk efficiently and effectively can be quite important.

The development of engineering has been a cumulative process and so it is the progress of risk engineering and management. The long history of engineering has left us a rich collection of examples, in terms of both successes and failures. In this book, a few selected examples have been the starting point of individual chapters. I hope that these examples will help the readers to understand some of the ways in which engineering systems can fail, the effects that these failures may impose, and the opportunities that engineers must turn failure into success.

While the focus of risk assessment as a discipline has traditionally been on potential failures and the associated negative consequences, this book introduces the term risk engineering. Risk engineering can be readily identified with the risks associated with costs and schedules on a project in which there is the potential for doing better as well as worse than expected. The ultimate goal of the discipline is to maximize engineering success. This is also the goal of every engineer, every engineering manager, and every engineering student.

The first edition of this book was published back in February 2000 and has received great audience. However, a lot of new information and material have taken place since then, primarily driven by artificial intelligence (AI), which is reshaping our lives, our jobs, and our world. Although we've known about AI for many years, its application in industry is still very much in the emergent stages. Seventy-five years ago, in 1947, Alan Turing said, "If a machine is expected to be infallible, it cannot also be intelligent". AI promises considerable economic benefits and significant technical challenges. For example, the self-driving car, as an important field of AI application, is progressing at an amazing pace due to the revolution introduced by deep learning technologies. However, AI also presents new and unique ways of failure. Unfortunately, some of the failures claimed people's lives as evidenced by the fatal crashes of cars with automated driving systems. Additionally, numerous research works have been published on how AI/neural network can fail because of hijacker's adversarial attacks.

It should be recognized that the topics of machine learning (ML) and AI are not just based on computer science; they also draw on computer engineering and a number of other disciplines. While the majority of computer science research focuses on solving software-related problems, such as how to use several programming languages, operating systems, and databases, the main goals of computer engineering are solving real-world engineering problems, including hardware and software interfaces. As a proof, quantum computing, quantum information, and quantum communication, which integrate new hardware and software infrastructures, are making great impacts to ML and AI. The broad perspective/mindset about ML and AI, necessary to the success of every engineering practitioner including engineers and engineering managers, needs to be addressed through What Every Engineer Should Know about (WEESKA) Series. This new edition provides an excellent venue to fulfill the need, since it is engineering bringing ML and AL to real world for practical applications for the benefits of the people.

It is important to recognize the risk of dangerously assuming "the issues are known in the field". Unknown unsafe scenarios are being addressed by Safety of the Intended Functionality (SOTIF), international standard ISO 21448:2022, just published in June 2022. Compliance with this international standard is especially important as AI and ML play key roles in the development of autonomous vehicles. According to the newly published international standard, it is crucial to identify and assess functional insufficiency of AI-based algorithms and the risks associated with their engineering implementations, for example, eye damage from the beam of a Light Detection and Ranging (LIDAR), a remote sensing method that uses light in the form of a pulsed laser to measure ranges (variable distances) to the Earth. Additionally, UL 4600: Standard for Safety for the Evaluation of Autonomous Products is the first safety standard for autonomous vehicle along with other applications and systems. UL 4600 is the first standard designed specifically for Autonomous, Automated and Connected Vehicles and related products, where AL/ML are playing important roles. Again, it is life-and-death to identify and assess functional insufficiency of AI-based algorithms and the risks associated with their engineering implementations. Dangerously assuming "all AI/ML largely sit in the computer science domain", rather than being part of the engineering one, is extremely risky as I emphasized above. Besides mechanical, civil engineering, so engineering where risk has a larger impact/importance, my new edition will be very beneficial to those whose job titles are related to computer science, computer engineering, and AL/ML engineering among many others such as aerospace engineering, aviation engineering, avionic engineering, safety engineering, reliability engineering, system engineering, mechanical engineering, manufacturing engineering, and industrial engineering. This new edition pioneers into the new horizon of Risk Engineering and Management.

Based on research work published by Taylor & Francis' International Journal of General Systems, it is the first text book introducing the term "Wang Entropy" that reveals complexity as a measure of the effort required to evaluate of a system; here, the system structure impacts system complexity significantly as well as the number of system components. Based on this core concept for Risk Engineering, this new edition further explores this new discipline with broad applications for Reliability, Availability, Maintainability, Testability, and Safety with emphasis on complexity evaluation of general systems in the age of AI.

I am excited about bringing the new addition to the WEEKSA series, to the readers around the Globe at a time when it is really needed. I have many persons to thank for their assistance in helping improve this new edition and want to especially thank Mr. Dennis M. Bushnell, HFAIAA, FASME, FRAeS (member of the National Academy of Engineering) for the Foreword of the new edition.

John X. Wang
PhD. CRE MBB

About the Author

Dr. John X. Wang has served as a Functional Safety Expert at Ford Motor Company and a Technical Director at Capgemini. As a Senior Principal Functional Safety Engineer at Flex, Dr. Wang has been spotlighted as the Employee of the Month by "Under the Hood", the online magazine published by Flex Automotive to employees around the Globe. Dr. Wang has authored/coauthored numerous books and papers on reliability engineering, risk engineering, engineering decision-making under uncertainty, robust design and Six Sigma, lean manufacturing, green electronics manufacturing, cellular manufacturing, and industrial design engineering – inventive problem-solving. As a recognized inventor, Dr. Wang has been granted two patents related to Risk Engineering by the US Patent and Trademark Office.

1 Risk Engineering
Dealing with System Complexity and Engineering Dynamics

1.1 UNDERSTANDING FAILURE IS CRITICAL TO ENGINEERING SUCCESS

The engineer is the driver of engineering design. To achieve engineering success, an engineer needs to understand and avoid potential failures that may occur down the road.

Engineering systems have their own "minds" and behavior, which may not be fully understood by their designers. A catastrophic bridge collapse may come from a lack of engineering knowledge about the structural characteristics or uncertainty regarding the load that the bridge will be required to bear 30 or 40 years after completion of the design. A catastrophic airplane crash may be caused by the growth of a tiny hidden flaw. A computer failure with a far-reaching impact may be caused by a minor "ill-logic" of its operating system.

Actually, every failure is a logical result of its causes, although properly diagnosing the cause may be difficult. Back in the 1830s, when engineers of steam engines drove the Industrial Revolution, the word *engineer* meant one who managed an engine. The sketch in Figure 1.1 shows the basic parts of a simple, single-cylinder steam engine. The term "valve gear" is not a traditional "gear", but instead it refers to the equipment that moves the slide valve.

The "Falcon" was an engine owned by the Galena & Chicago Union Railroad and was intended for heavy passenger service of the time. It was very solidly built with cylinders 15×22 inches and two pairs of driving wheels 66 inches in diameter. A group of Chicago engineers undertook to design this engine so it would be better than a McQueen engine, their competitor's.

When the eventful day of its public unveiling arrived, the Falcon was coupled onto 16 passenger cars and pulled out in the midst of an ovation. The operators maximized the steam pressure to get all of the power possible. It was noted that they had overweighed the safety valves and the crownsheet was bagged between the stay-bolts through the effects of excessive pressure. Even with the recklessly high pressure, the trip to Freeport was terribly long.

On the next day, a McQueen engine, of the same size as the Falcon, coupled onto seven passenger cars and pulled out. This train made all the stops and reached Freeport five minutes ahead of time. The Falcon was beaten. It was later tried on all

DOI: 10.1201/9781003371014-1

FIGURE 1.1 Steam engine of similar vintage to the Falcon.

kinds of service but in every case failed to match the performance of other engines of the same class. The underlying or root cause of this performance failure was not understood. It was simply assumed that the design was bad.

The Falcon was apparently the kind of bird that could not fly, run, or pull. Most of the engineers gave up the engine as a hopeless case. However, this opinion was not shared by Wm. Wilson in the Galena shops in Chicago. One day when the engine was in for repair, he obtained permission to do a thorough inspection.

The poor performance of the Falcon could be due to something inhibiting the steam flow. The engineer first went over the valve motion and found it to be in about the same condition as other engines. Then, he took the dome cap off and found that nothing obstructed the steam entry into the dry pipe. He followed with an inspection of the steam passages between the steam chest and cylinders and found nothing blocking the steam there.

The case was very puzzling, but Mr. Wilson had infinite faith in steam acting the same in this engine as it did in others, if permitted to reach the cylinders. Therefore, he decided to continue to follow the passages back to the throttle valve. As a preliminary, he took down the steam pipes and found that they were in about the same condition as other locomotives. When he got up to the branch pipe to feel inside the dry pipe, he found an obstruction!

The branch pipe was taken down, and it was discovered that a centering plate, which had been inserted into the dry pipe when the joint was being turned, had not been removed. The only opening left for the steam to pass through had been two small holes that were in the plate. *When the plate was removed, the Falcon proved to be as good an engine as any of its competitors on the road.*

To drive an engineering complex, engineers need to foresee the possibility of failures like that experienced by Mr. Wilson. This foresight is vitally important to find the success path to a good engineering product. Currently, many risk engineering tools, such as Fault Tree Analysis (see Chapter 3) and Failure Mode and Effect Analysis (see Chapter 4), are available to help us foresee potential failures and their impact. Fault-tolerant design (see Chapter 4) is widely used in today's electric or diesel-electric locomotives.

Engineers must get to the root causes of failures and through this understanding ensure a clear path to engineering success.

1.2 RISK ASSESSMENT – QUANTIFICATION OF POTENTIAL FAILURES

If we cannot express what we know in the form of numbers, we really don't know much about it. If we don't know much about it, we cannot expect to optimally control it. That is why we need to quantify potential failures.

Risk assessment is the quantification of potential failure and needs the answers to the following three questions:

1. What can go wrong within an engineering system?
2. How likely is the failure to happen?
3. What will be caused by the failure as a consequence?

Engineers begin their design endeavor by answering "what might work and what can go wrong". Engineering designs are successful only to the extent that designers foresee and understand how a system may fail to perform its necessary functions. When designing a bridge, the engineer needs to know what loads could challenge the strength of the individual steel members. When designing a communication system, the engineer needs to predict what might cause connection problems within the communication network. When designing a steam boiler, the engineer needs to anticipate, for example, the failure scenarios that would build up pressure and cause a boiler explosion.

The concept of risk combines chance for failure with the consequence caused by the failure. An essential element of risk is the uncertainty – the fact that engineers don't know exactly what failures will occur and when and where the failures will occur. The assessment of risk involves both probability and consequences, which will be detailed in Chapter 3.

Example 1.1

For this example, we consider an energy system consisting of the following three units:

- Energy Generation Unit (EGU) – Generate Energy from Fuel Inside;
- Energy Control Unit (ECU) – Control Energy Generation According to Requirements;
- Energy Transport Unit (ETU) – Transport Energy to Distribution Network.

TABLE 1.1
Risk of the Energy System (Example 1.1)

What Can Go Wrong?	How Likely Is the Failure to Happen? P_i	What Is the Consequence (Cost Per Accident)? C_i
EGU Generates Too Little Power	1 Time Per Year	Does Not Satisfy Customer Requirements ($5,000 Per Accident).
EGU Generates Too Much Power to be Controlled by ECU	0.005 Times Per Year	Energy System Damage Caused by Explosion due to Accumulated Energy ($1,000,000 Per Accident).
ECU Fails to Control Power Output Accurately	2 Times Per Year	Does Not Satisfy Customer Requirements ($5,000 Per Accident).
ECU Completely Loses Control	0.001 Times Per Year	Catastrophic Failure Causes System Damage and Possible Casualties ($10,000,000 Per Accident).
ETU Fails to Transport Energy to the Network; Pollution released to Environment Instead	0.02 Times Per Year	Environmental Impact ($500,000 Per Accident).
Loss of ETU	0.005 Times Per Year	Energy System Damage Caused by Explosion due to Accumulated Energy ($1,000,000 Per Accident).

We will quantify the risk of operating this system, combining the risk contributions from all potential accidents. The risk triplets of the energy system are identified in Table 1.1 along with a listing of their frequency. The total risk is calculated as follows:

$$Risk = \sum_i P_i C_i$$

Therefore,

$$\begin{aligned} \text{Risk} &= 1\times 5,000 + 0.005\times 1,000,000 + 2\times 5,000 \\ &\quad + 0.001\times 10,000,000 + 0.02\times 500,000 + 0.005\times 1,000,000 \\ &= \$45,000 \text{ per year} \end{aligned}$$

Thus, operators of such systems will realize that some years will pass in which there are no accidents, but on average over many similar systems operated for many years, the cost of accidents will be $45,000 per year. This provides a quantitative measure of risk associated with system operation.

The process of risk assessment could be used by an insurance company that wishes to insure the operating company against potential losses due to these particular sources of risk. Risk engineers for the operating company could evaluate the benefit of certain changes in the system which would reduce or eliminate selected risk contributors and thereby reduce the cost of the insurance. This risk engineering approach as a part of risk management has the potential for positive benefits.

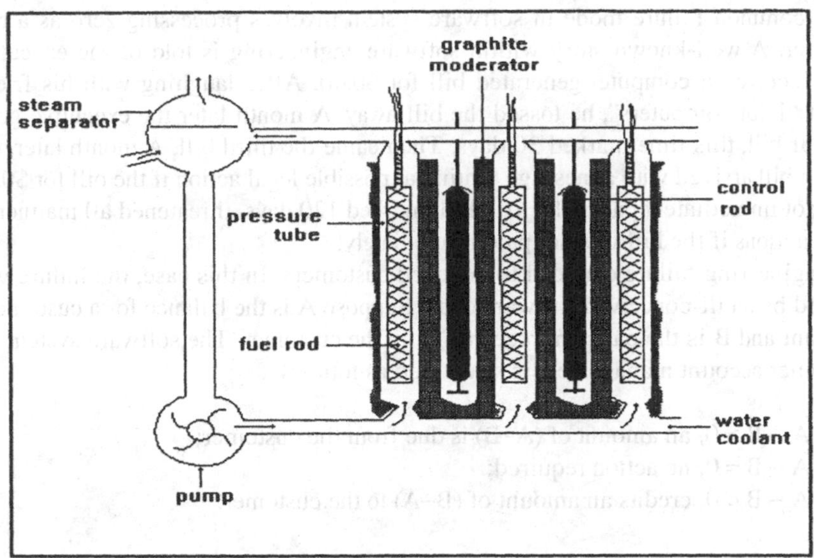

FIGURE 1.2 Reactor core of Chernobyl Unit 4 (Boiling Water Reactor).

Energy is a valuable asset if under control, but it may pose an extreme hazard if out of control. As shown in Figure 1.2, reactor number 4 of the Chernobyl Nuclear Power Plant in Ukraine was a huge energy complex. The energy generation was maintained by fission reactions inside the reactor core. Energy was transported from the reactor core to the outside heat exchanger through a water-cooling system. The nuclear power reactor was controlled by 211 control rods, neutron-absorbing materials, also in the reactor core.

People have a good idea what can happen if a car's brakes are disabled. For a nuclear reactor, the chain reaction will grow out of control if too many control rods are pulled out. On the early morning of April 26, 1986, most of the control rods in the Chernobyl Reactor 4 were pulled out of the reactor core – the energy giant was running in a very unstable condition without brakes! Everyone now knows of the appalling disaster which took place that early morning.

The operation of nuclear reactors is commonly viewed as a very risky business because of the dire consequences possible with catastrophic accidents. Because some other systems have higher frequencies of accidents, they may represent an equal risk to the operator even though the consequences of these accidents are much smaller. The total risk of operating a system must include the low-consequence events along with the high-consequence events. The risk assessment of a nuclear power reactor using an event tree analysis will be detailed in Chapter 3.

1.3 RISK ENGINEERING – CONVERTING RISK INTO OPPORTUNITIES

A concept common to both hardware and software products is the concept of risk, the expected consequence due to engineering failures. Risk and failures are associated with operating software systems as well as hardware systems. Risks are present wherever we go.

A common failure mode in software system involves processing zero as a real number. A well-known story within software engineering is told of the executive who received a computer-generated bill for $0.00. After laughing with his friend about "idiot computers", he tossed the bill away. A month later the executive got a similar bill, this time marked 30 days. Then came the third bill. A month later, the fourth bill arrived with a message hinting at possible legal action if the bill for $0.00 was not immediately paid. The fifth bill, marked 120 days, threatened all manner of legal actions if the bill was not paid immediately!

Engineering failure can lead to annoyed customers. In this case, the failure was caused by an ill-conceived software code. Suppose A is the balance for a customer's account and B is the payment received from the customer. The software system for customer account management is structured as follows:

If A – B > 0, an amount of (A–B) is due from the customer;
If A – B = 0, no action required;
If A – B < 0, credits an amount of (B–A) to the customer.

The structure seems perfect. However, experienced software engineers will indicate the required conversion from decimal number to binary number. This ($32.56–$32.56) may produce a very small number (e.g., 0.000008) rather than an exact zero. Also, the amount due may be a calculated number displayed with two-digit decimal point accuracy, while the amount paid is precisely a two-digit number. The generated positive number, although it is very small, led to a computer-generated bill for $0.00, which was actually a two-decimal representation of $0.000008.

Figure 1.3 shows an irate executive is smashing his "idiot computer" with a hammer. Fearful of the loss of his organization's credit rating in the hands of this threatening machine, the executive called a friend who was a software engineer and related the whole sorry story. The software engineer suggested that the executive mail in a check for $0.00 attached with the following suggested a software structure for the account management system:

If ABS(A–B) # δ (δ is a small positive real number), then no action required;

Otherwise,

IF A – B > δ, an amount of (A–B) is due from customer;
Otherwise, credit an amount of (B–A) to the customer.

Like any engineering design, a software system with no tolerance specification often leads to wrong results, which will, in turn, impact customers' satisfaction. As illustrated by the story above, to design reliable software, it is important to understand how failures may be introduced and how to avoid them. Software is now operating a wide range of products and systems, and this trend is accelerating with the opportunities presented by low-cost microprocessors. However, residual faults ("bugs") may hide within any software system. Such a bug delayed the first space shuttle orbital flight.

On April 10, 1981, about 20 minutes before the scheduled launching of the first flight of the Space Shuttle, astronauts and technicians attempted but failed to initialize

FIGURE 1.3 Executive and the "idiot computer".

the software system that backs up the four-fold-redundant primary software systems. It turned out that the Backup Flight Control System in the fifth onboard computer could not be initialized with the Primary Avionics Software System already executing in the other four computers, due to a small error in the software. The result was a very costly delay in the launch. For customers and developers, budget overruns and schedule slip are part of the risk equation.

Example 1.2

A fundamental risk-analysis paradigm is the decision tree, which is the basis of the event tree analysis in Chapter 3. Figure 1.4 illustrates such a decision tree for a potentially risky situation involving the software controlling a satellite experiment. The software has been under development by an engineering team, which understands the experiment but is unfamiliar with the best software engineering practice. As a result, the satellite-experiment project manager has obtained an estimate that there is a probability of 0.4 that the experiments' software will have a critical error, which will destroy the entire experiment and result in a loss of the total $20 million investment in the experiment.

The satellite-experiment project manager identifies two major options for reducing the risk of losing the experiment.

- Directing the engineering team to utilize better software development methods. The training of the engineering team would incur an additional cost of $10,000. The manager estimates that this will reduce the failure probability to 0.1.

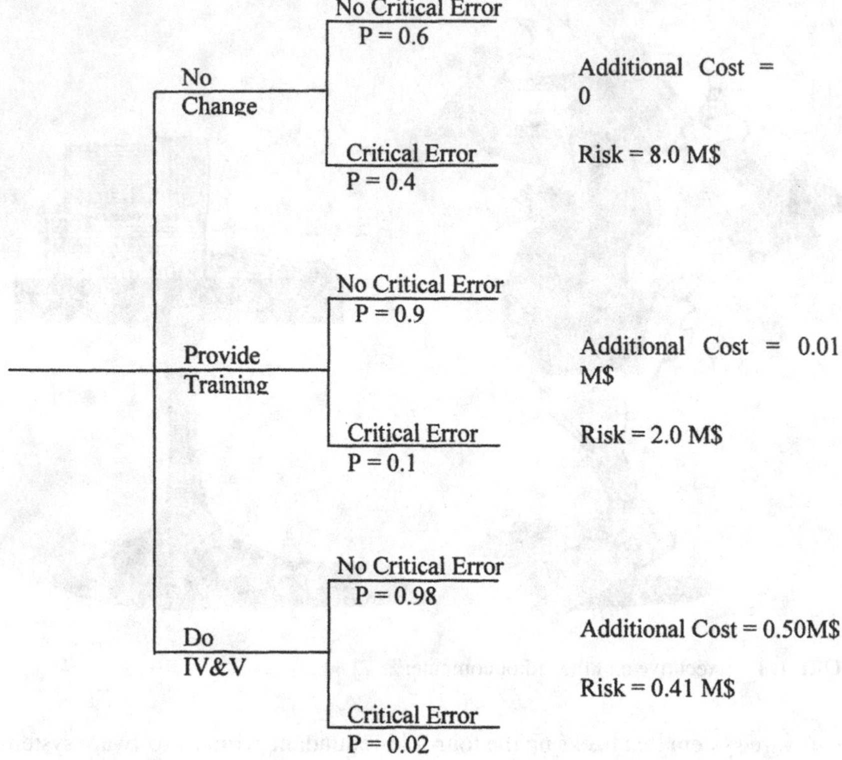

FIGURE 1.4 Decision tree for risk associated with alternative strategies.

- Hiring a contractor to independently verify and validate the software. This costs an additional $500,000; based on the results of similar IV&V efforts, the manager estimates that this will reduce the failure probability to 0.02.

For each of the three major decision options, the decision tree in Figure 1.4 shows the possible outcomes in terms of the critical error remaining, their probabilities, the cost, and the risk exposure associated with each outcome.

The expected risk exposure with no change made is

$$\text{ER}(\text{Choice}\,\#1) = 0.40 \times 20 = 8.0 \text{ million dollars}$$

With the use of better software development methods, there is an additional fixed cost of $0.01 million and the expected risk exposure is

$$\text{ER}(\text{Choice}\,\#2) = 0.1 \times 20 = 2.0 \text{ million dollars}$$

If independent verification and validation (IV&V) is done, there will be an additional fixed cost of $0.5 million and the expected risk exposure is

$$\text{ER}(\text{Choice}\,\#3) = 0.02 \times 20 = 0.4 \text{ million dollars}$$

With today's shrinking budgets and increased workloads, it is imperative to introduce practical, logical, systematic, and economic methods of software risk reduction and quality assurance into the development life cycle. These methods must recognize how software fails, the causes, and the need for risk reduction, mitigation, and control techniques that function smoothly within the engineering disciplines of the organization. In Chapter 2, we will further discuss the various failure modes during software development.

Risk implies opportunities for improvement. Success in engineering benefits from the ability to recognize opportunities.

Thomas Edison's idea for a light bulb filament was inspired by the failure in which a filament broke. By analyzing the electrical resistance and power supply network, he recognized the opportunity to improve the light bulb design with new filament materials. He tested thousands of potential materials before coming up with a satisfactory light bulb design. Validation and qualification test is an important part of risk engineering.

Compared with light bulb filament breakage, the consequences of breaks in airplane structures are more significant.

The farther we are away from the Earth, the more at risk we feel. In the 1930s, engineers began to recognize the opportunity offered by higher altitude – flying faster with turbojet engines. Figure 1.5 shows a Comet jet airplane. The first commercial jet airplane, the Comet, was placed in service on May 2, 1952. However, this was only the beginning of managing the risk presented in high-altitude manned flights.

The early flights of the Comets suffered a series of fatal accidents. On May 2, 1953, the first anniversary of jet airplane service, a Comet exploded in midair while taking off from Calcutta, India. On January 10, 1954, another Comet exploded at 27,000 feet near the island of Elba in the Mediterranean Sea. Then, on April 8, 1954, the third midair explosion of a Comet occurred when the airplane was taking off from Rome. There was no evidence of explosives involved in these airplane accidents.

Like Thomas Edison, engineers resorted to thorough investigations in searching for the causes of these midair explosions. They immersed a retired Comet in a tank

FIGURE 1.5 A Comet jet airplane.

of water. To simulate the repeated pressurization and depressurization during the flight mission, water was alternately pumped into and out of the cabin. The wings were simultaneously flexed by hydraulic jacks to represent the forces of the air during flight. The test was repeated for about 3,000 simulated flights when, suddenly, the comer of one of the cabin windows developed a crack. Growing rapidly under continued flight simulation, the crack finally spread catastrophically through the plane's metal skin!

After fixing the design problem with reinforced window panels, a new model Comet 4 initiated trans-Atlantic jet passenger service in 1958. The name of Comet 4 could mean that its success came from the three previous failures. This ultimate engineering success pioneered the age of international aviation and brought tremendous opportunities for aerospace engineering. However, risk acceptance is by no means a simple topic, especially when there is a potential risk of human casualties. In Chapter 5, we will show how to build risk acceptance on a sound engineering basis.

Each of these examples gives a hint of the pervasiveness of risks presented by every system or major undertaking. The elements of risk engineering will be discussed thoroughly in this book and guidance provided for how we can manage risk.

1.4 SYSTEM COMPLEXITY – MEASURED BY WANG ENTROPY

This section investigates complexity as a measure of the difficulty of diagnosing or troubleshooting a system. It has been discovered that system complexity and the number of system components have an impact on the system structure. Wang Entropy is defined as the entropy of minimal cut set importance, which can be utilized to measure system complexity.

1.4.1 Imagination for Problem-Solving: Transfer a Risk Engineering Problem into a Thermodynamic Problem

Imagination is important for inventive problem-solving. How to transfer a risk engineering problem into a thermodynamic problem, thus leverage the Second Law of Thermodynamics "Entropy"?

There is a small link between a risk engineering problem and a thermodynamic problem. As shown in Figure 1.6, if we label the portion left to the separator as

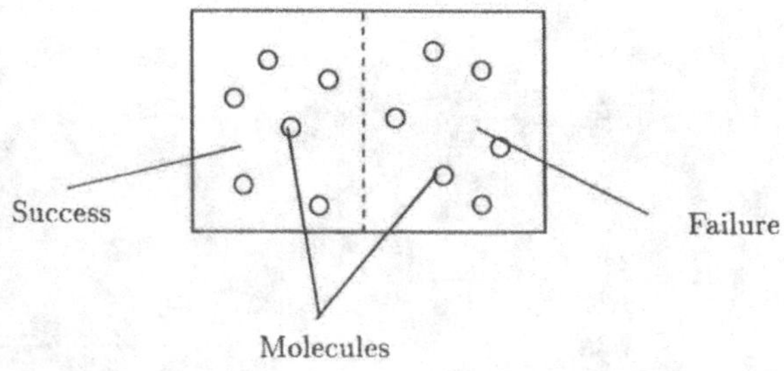

FIGURE 1.6 Transfer an N-Molecules Thermal System into an N-Components System for Risk Engineering.

"Success", and the portion right to the separator as "Failure", we can transfer an N-Molecules Thermal System into an N-Components System for Risk Engineering.

However, an engineering system is different from a thermodynamics system, in that an engineering system has a specific structure. Besides the number of components and the probabilistic distribution of system states, the complexity of a system also depends on its structure. In the next section, Wang Entropy as a measure of system complexity will reflect the impact of the system's structure on its complexity.

1.4.2 Wang Entropy as a Measure of System Complexity

Complexity plays a very important role in systems problem-solving. It emerges as a fundamental concept of systems science, one that is perhaps as fundamental as the concept of energy is to natural science. The concept of complexity has many specific meanings, such as computational complexity, cognitive complexity, and system complexity. Systems science challenges itself to capture the general characteristics of this important concept.

There is an interesting link among the concepts of information, uncertainty, and complexity. The uncertainty-based complexity is proportional to the amount of information needed to resolve any uncertainty associated with the system involved. In this section, we explore the concept of complexity from the perspective of system reliability and reveal the following two points:

- Uncertainty: A special uncertainty associated with an engineering system is manifested in the level of difficulties necessary to find failures within the system.
- Information: Information needed to resolve the above uncertainty can be measured by the minimum number of average inspections necessary to isolate system failures.

System engineers must translate general system characteristics, including reliability, into detailed specifications for the numerous units that make up the system. The process of assigning reliability requirements to individual units to attain the desired system reliability is known as reliability allocation. Reliability is usually apportioned on the basis of unit importance and complexity. The unit complexity is defined in terms of modules and their associated circuit, where a module is an electron tube, a transistor, or a magnetic amplifier. Using the number of components as a measure of system complexity is only applicable to series systems consisting of the same components. But how should we define system complexity for a general system?

Besides the number of components and the probabilistic distribution of system states, the complexity of a system also depends on its structure. For example, as shown in Figure 1.7, four components can be organized into many systems. The complexities of these systems are quite different.

Cut set and minimal cut set are important concepts for a system reliability analysis. A cut set is a set of units that interrupt all possible connections between the input and output points. A minimal cut set is the smallest set of units needed to guarantee an interruption of flow. A component or cut set's contribution to the system failure

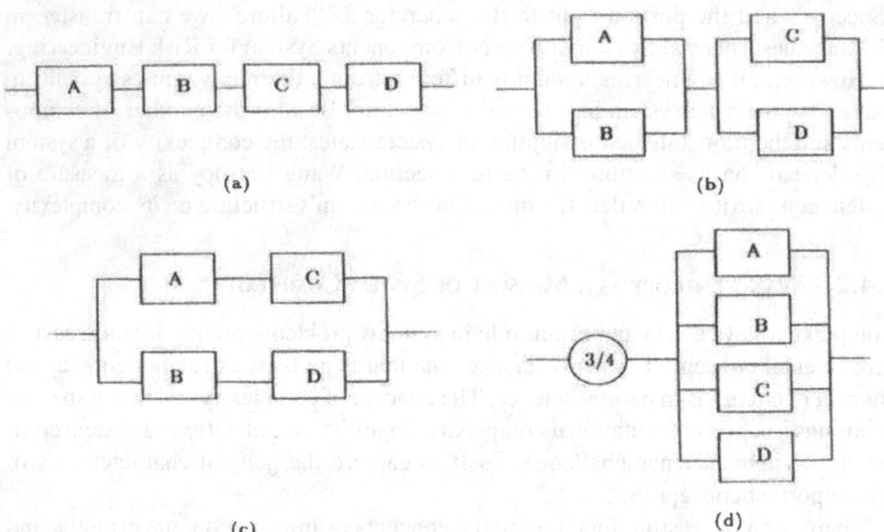

FIGURE 1.7 Different systems structure comprising four components (a–d).

is termed its importance. The specific structure of an engineering system provides the connection between Wang Entropy, the entropy of cut set importance, and the complexity of system diagnosis. The nature of this intrinsic relationship will be discussed below.

Recent research work on system fault diagnosis reveals that Wang Entropy, a system-intrinsic feature, presents a lower bound to the average number of inspections to find the failure mode leading to a system failure. This is very similar to medical diagnosis. When diagnosing a disease, physician A needs 20 inspections to find the disease cause; physician B needs 14 inspections, while physician C needs only 10 inspections to find the disease cause. C is, therefore, the best physician among the three physicians. However, the average number of inspections to find the disease has a lower bound which is an inherent feature of the diagnosis problem. This lower bound presents a measure of the difficulty of the diagnosis problem. Some diseases are difficult to diagnose because they need many inspections to find their cause. In system fault diagnosis, the system-intrinsic feature which gives the lower bound to the average number of inspections to find the failure mode can be used as a measure of uncertainty in fault diagnosis.

While the failure modes for a system can be expressed by its Minimal Cul SeL (MCS), the average number of inspections to find the minimal cut set causing a system failure is found dependent on the inspection sequence adopted. However, this average number of inspections is proven to be lower bounded by Wang Entropy, which may be used to estimate how difficult it is to find the actual MCS. This entropy function presents an intrinsic feature of the system and can be used as a measure for system complexity, which is significant to reliability prediction and allocation. This intrinsic feature measures how difficult it is to diagnose the system and can be used as a measure for system complexity.

Cut set importance: It is the MCS occurrence probability, given that a system has failed. If the prior probabilities of MCSs are $P(C_1)$, $P(C_2)$, $P(C_n)$, then the importance of cut set C can be evaluated by

$$I_i = \frac{P(C_i)}{\sum_{i=1}^{n} P(C_i)}$$

Wang Entropy: It is defined as the entropy of minimal cut set importance, which can be utilized to measure system complexity. Wang Entropy can be calculated as follows:

$$W_H = \sum_{i=1}^{n} I_i log_2$$

It is shown in this section that Wang Entropy gives a lower bound to the average number of inspections which must be conducted in any inspection sequences designed to determine the actual MCS. This entropy function, which presents an intrinsic feature to measure the level of difficulty in finding the failure mode causing system failure, can be used to measure the system complexity. The reliability of complex equipment depends on the system structure and the reliability of the assemblies and components which comprise it. The equipment reliability can be allocated among its elements based on Wang Entropy, as a measure of system complexity, discussed in this section. For a system composed of four components, Table 1.2 shows the effect of system structure on system complexity.

1.5 ENGINEERING – A PROFESSION OF MANAGING TECHNICAL RISK

Engineers deal daily with risk, both professionally in providing advice to then-clients and commercially in operating their own business, whether self-employed or as an employee. The engineering risks are increasing as projects become more complex. In response to this, risks need to be identified, evaluated, and managed in a formal system of control rather than the informal systems which have existed in the past.

The primary responsibility of an engineer is to make technical decisions. Often, such decisions have to be based on incomplete information; thus, the outcomes of engineering decisions invariably involve uncertainty. In these circumstances, risk is inherent in the engineering decision-making process.

For engineering projects, risk is a measure of the uncertainty of the project outcomes. Broadly speaking, risk can be defined as the deviation of project outcomes from a mean or anticipated value. It can also be regarded as the chance of incurring a loss or gain by investing in an engineering project. The chance of making a profit or incurring a loss can be high or low depending on the risk (variability of possible outcomes) associated with a given project.

TABLE 1.2
The Effect of System Structure on System Complexity

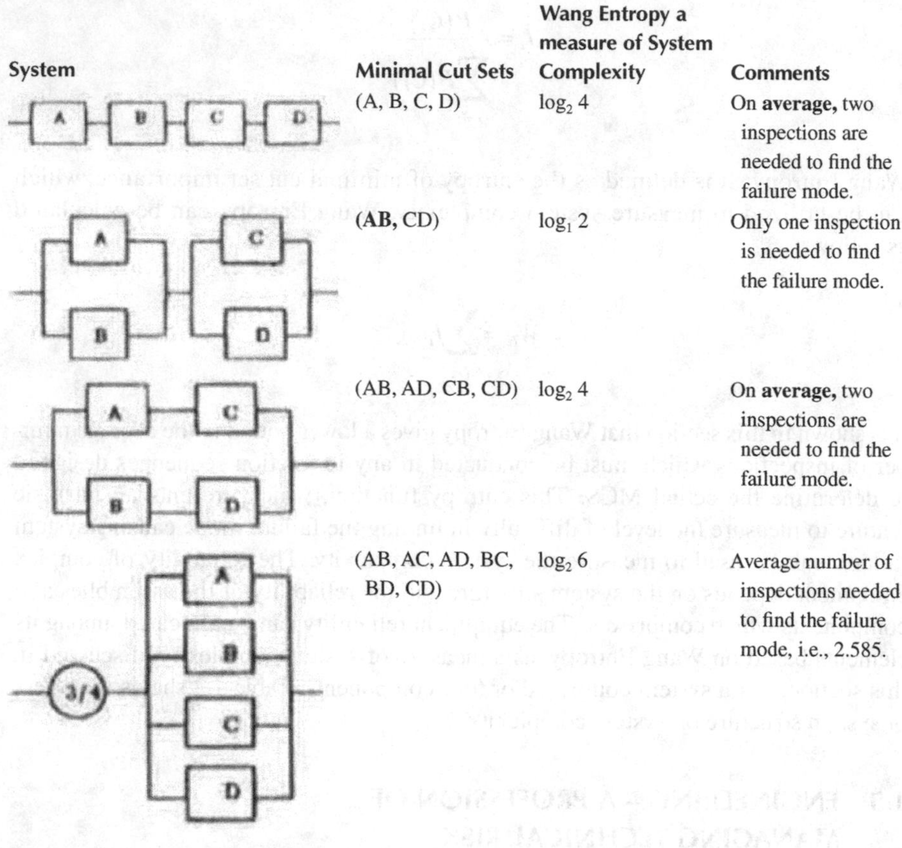

System	Minimal Cut Sets	Wang Entropy a measure of System Complexity	Comments
A–B–C–D (series)	(A, B, C, D)	$\log_2 4$	On **average,** two inspections are needed to find the failure mode.
(A ∥ B)–(C ∥ D)	(**AB**, CD)	$\log_1 2$	Only one inspection is needed to find the failure mode.
(A–C) ∥ (B–D)	(AB, AD, CB, CD)	$\log_2 4$	On **average,** two inspections are needed to find the failure mode.
3/4 (A, B, C, D)	(AB, AC, AD, BC, BD, CD)	$\log_2 6$	Average number of inspections needed to find the failure mode, i.e., 2.585.

Probabilities help us to determine the likelihood or chance of an event occurring. Some probabilities can be obtained from actual observations. For example, the risk of getting heads or tails from a coin can be readily measured by flipping it many times and finding out the actual outcomes. However, there are instances – such as the introduction of a new product – when the outcome is highly uncertain. In these cases, there is usually little or no past experience to draw on. The engineer must make a judgment as to the probable outcome.

Example 1.3

Two alternative designs are considered for a Magnetic Resonance Imaging (MRI) system. Design A is based on a conventional instrument, whose reliability of satisfactory performance is 95%, and costs $2.5 million. Design B is based on innovative concepts and will reduce the cost of building the system by $1 million. The

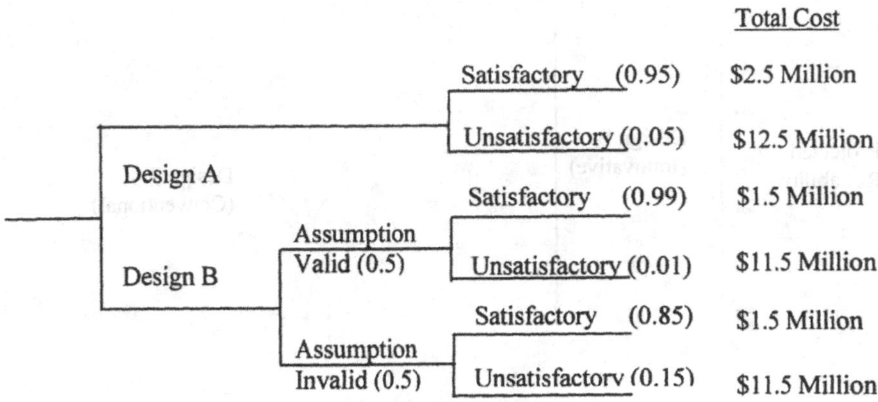

FIGURE 1.8 Decision tree for an MRI system design.

reliability of design B is not known; however, the engineer estimates that if his assumptions are correct, the reliability of satisfactory performance will be 0.99; however, if his assumptions are not valid (equally probable), the reliability will be only 0.85. Suppose the liability cost of unsatisfactory performance is $10 million. The engineer now faces a decision between the following two alternatives:

- Capture the opportunity of accomplishing cost-saving and high reliability by the innovative design; or
- Assure a reasonable reliability by the conventional design.

A decision tree is often used to help engineers make technical decisions. Figure 1.8 shows the decision tree for use in assisting with the selection of the MRI system design.

The expected total cost for design A can be calculated as follows:

$$\text{Expected Total Cost (A)} = 0.95 \times 2.5 + 0.05 \times 12.5 = \$3 \text{ (Million)}$$

The expected total cost for design B can be calculated as follows:

$$\text{Expected Total Cost (B)} = 0.5 \times (0.99 \times 1.5 + 0.01 \times 11.5) + 0.5 \times (0.85 \times 1.5 + 0.15 \times 11.5) = \$2.3 \text{ (Million)}$$

Regarding the expected total cost, design B (even with its uncertainties) is a reasonable choice. Risk engineering provides a systematic basis for evaluating the probabilities and risk associated with each branch of the decision tree.

Because the future outcome from a new product is often highly uncertain, an engineer will work with the assumption that the projected outcome will fall within a particular range. The more uncertain the project outcome, the bigger the range. For the MRI system design in Example 1.3, the reliability of the conventional design, design A, is large (95%) based on extensive historic data; however, the reliability performance with the innovative design, design B, is quite uncertain due to the dependency on engineering assumptions; its reliability is estimated to be between 85% and 99% (see Figure 1.9).

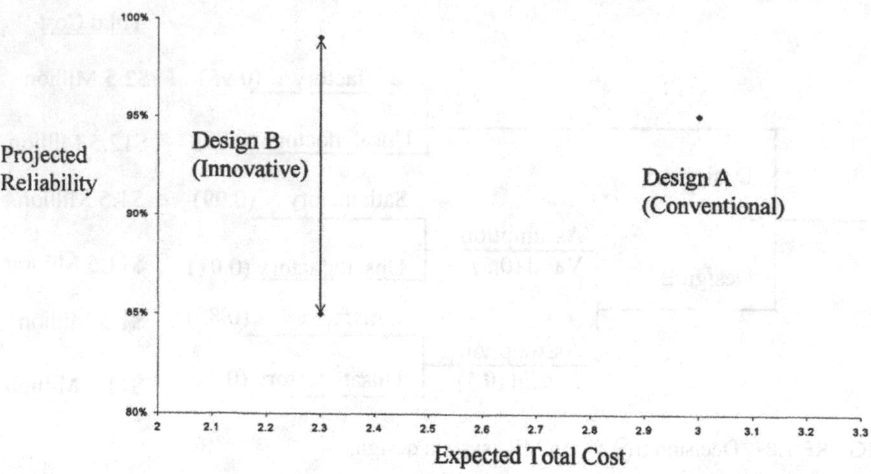

FIGURE 1.9 Conventional design vs. innovative design.

To establish a sound engineering project, it is important to understand the risks and opportunities provided by the project. Essentially, you must answer two inseparable questions:

- What is my project objective?
- How much uncertainty can I tolerate as I seek my project objective?

For engineering projects, there's always a trade-off between risk and opportunity. For the problem in Example 1.3, the innovative design provides the opportunity for achieving cost-savings and improving reliability but it also presents a considerable risk, the uncertain reliability performance. A conservative engineer might select the conventional design to assure a reasonable reliability, while an aggressive engineer might take the innovative design to achieve better project performance.

Risk engineering is an integrated process which includes the following two major parts:

1. Through risk assessment, uncertainties will be modeled and assessed, and their effects on a given decision evaluated systematically;
2. Through design for risk engineering, the risk associated with each decision alternative may be delineated and, if cost-effective, measures are taken to control or minimize the corresponding possible consequences.

The following chapters will illustrate the basic elements of risk engineering.

Chapter 2 Risk Identification – Understanding the Limits of Engineering Designs
Every engineering design has its limitations and its breaking point. By recognizing all the possible mechanisms of failure, a robust engineering design can be developed to minimize the potential risk. Risk identification starts by describing the system structure and how it interfaces with operating environments.

Chapter 3 Risk Assessment – Extending Murphy's Law
Murphy's Law tells us, "If anything can go wrong, it will". Risk analysis provides us with an additional insight about how likely it is to go wrong,

and what it will cause as a consequence. Risk analysis will provide the quantitative inputs for engineering design.

Chapter 4 Design for Risk Engineering – The Art of War Against Failures

Design for Risk Engineering starts from the weakest link of an engineering system, and is built on the following Three Lines of Defense:

- The First Line of Defense – Avoid or Eliminate Failure Cause;
- The Second Line of Defense – Detect and Control Failure Early;
- The Third Line of Defense – Reduce Impact/Consequence of Failures.

Chapter 5 Risk Acceptability – Uncertainty in Perspective

Uncertainty is inherent in material strength, engineering design, manufacturing processes, and operating environments. Acceptable risk can be accomplished only through the capability of controlling uncertainties associated with each phase of the engineering life cycle.

The second part (Chapter 6 to Chapter 9) of this book steps back to provide a broad view of integrated risk management for technical risk, cost risk, and schedule risk. Engineers always challenge the boundaries of established engineering and technologies, particularly where they find risk and opportunity.

BIBLIOGRAPHY

AT&T and the Department of the Navy. (1993), *Design to Reduce Technical Risk*, McGraw-Hill Inc., New York.

Bengtsson, G. (1989), "Risk Analysis and Safety Rationale," Nordic Liaison Committee for Atomic Energy.

Carnegie Commission on Science. (1993), *Risk and the Environment – Improving Regulatory Decision Making*, Technology and Government, New York.

Littlewood, B. and Strigini, L. (1992), "The Risks of Software," *Scientific American*, Vol. 267, No. 5, pp. 62–75.

McCarty, L. S. and Power, M. (1996), The Role of Science in Risk Management Decision Making, SETAC Annual Conference.

Miccolis, J. A. (1996), "Toward a Universal Language of Risk," *Risk Management*, Vol. 43, pp. 45–49.

National Research Council. (1983), *Risk Assessment in the Federal Government: Managing the Process*, National Academy Press, Washington.

Petroski, H. (1985), *To Engineer Is Human – The Role of Failure in Successful Design*, St. Martin's Press, New York.

Roush, M. L., Modarres, M., and Hunt, R. N. (1985), "Application of Goal Trees to Evaluation of the Impact of Information upon Plant Availability," Proceedings of the International ANS/ENS Topical Meeting on Probabilistic Safety Methods and Applications, San Francisco.

Royal Society. (1992), *Risk Analysis, Perception and Management*, The Royal Society, London.

Wang, J. X. (1996), "Complexity as a Measure of the Difficulty of System Diagnosis," *International Journal of General Systems*, Vol. 24, No. 3, pp. 257–269.

Wang, J. X. (2017), *Industrial Design Engineering: Inventive Problem Solving*, CRC Press, Boca Raton, FL.

Wang, J. X. (2019a), "Complexity as a Measure of the Difficulty of System Diagnosis in Next Generation Aircraft Health Monitoring System," SAE Technical Paper 2019-01-1357, doi:10.4271/2019-01-1357.

Wang, J. X. (2019b), "A Dynamic Fault Tree Approach for Time-Dependent Logical Modeling of Autonomous Flight Systems," SAE Technical Paper 2019-01-1358, doi:10.4271/2019-01-1358.

2 Risk Identification
Understanding the Limits of Engineering Designs

2.1 THE FALL OF ICARUS – UNDERSTANDING THE LIMITS OF ENGINEERING DESIGN

Daedalus, of Greek legend, was known throughout the land for his ingenuity and craft. He was a sculptor, architect, and inventor without peer. He can be thought of as the first aeronautical engineer because of his mythical wing-making.

When the engineer and his young son Icarus were imprisoned in a high tower by the powerful King Minos, who controlled the land and the sea, Daedalus formed an engineering design for flying through the air. Innocently requesting candles so that he might continue to read and study, and using their wax and the feathers of birds that flew about the tower, he constructed a pair of wings.

Carefully observing the birds, Daedalus placed large feathers over small so as to form an increasing surface. He secured the larger ones with thread and the smaller ones with wax, and gave the whole a gentle curvature just like the wings of the birds. When at last the work was done, the craftsman waved the artificial wings and to his delight found himself hovering at the ceiling of his cell. He was flying!

At a much faster pace, he now constructed a smaller set of wings for his son. As shown by Figure 2.1, after the two practiced for a while, Daedalus carefully instructed Icarus. *Stay at a moderate height. Fly too low and the dampness of the sea will clog your wings. Fly too high and the heat of the sun will melt them.*

And with that, out through the window they flew. Along the shore one after another stared up at the pair in amazement. Surely, these were gods in the form of humans, sailing through the sky.

Icarus became entranced with the joy and power of flight and forgot his father's cautions. He tilted his wings and soared upward as if to the heavens. Soon the blazing sun softened the wax that held the feathers together, and they began to drop off. Icarus noticed that he was descending and fluttered his arms faster and faster but to no avail. As the father watched in horror, his son plunged into the sea. Daedalus circled again and again over the spot where the boy had gone down, but nothing rose to the surface except a handful of feathers.

Every engineering design has its limitations and its breaking point, just as Daedalus' wing-design could be fouled by the water or melted by the sun. Table 2.1 lists major aircraft accidents since the 1970s. Engineers must understand the limits of their design to avoid another fall of Icarus.

DOI: 10.1201/9781003371014-2

FIGURE 2.1 Design limit: Not too close to the sun!

2.2 OVERLOAD OF FAILURES: FRACTURE AND ITS MECHANICS

Figure 2.2 shows "London Bridge Is Falling Down", a traditional English nursery rhyme and singing game. London Bridge was really falling down. In 1962, it was discovered that the bridge, which was constructed in the 1820s, was inadequate for modern London's traffic and was slowly sinking into the River Thames.

Overload leads to fracture and other failure modes as summarized in Table 2.2. Mechanical loads may be in the form of tension, compression, or shear. Bending loads cause tensile and compressive forces, but fracture usually occurs in tension. Fracture can be either ductile, occurring after plastic deformation of the material, or brittle,

> Ductile fracture results from relatively slow crack growth rates, with high energy needed to continue the process;
> Brittle fracture results from very fast crack propagation in materials, with low energy required to initiate and sustain it.

Positive aspects of the phenomenon of brittle fracture behavior have influenced the quality of life from the beginning of human existence. Prehistoric man used brittle fracture to make stone tools.

The fact that many structures that the Egyptians and Romans built are still standing is testimony to the ability of early architects and engineers. The ancient structures that are still standing today obviously represent very successful designs. The Romans supposedly tested each new bridge by requiring the design engineer to stand underneath while chariots drove over it. That will separate the good from the bad!

TABLE 2.1
The Fall of Modem Icarus (Major Aircraft Accidents since the 1970s)

Aircraft Accidents	Accident Description
SwissAir 111, Halifax, Nova Scotia, Canada (September 1998)	A pilot reported smoke in the cockpit, dumped tons of fuel, and attempted an emergency landing before his MD-11 jetliner crashed into the ocean off Nova Scotia, killing all 229 people aboard, including 136 Americans. Flight 111 from New York to Geneva plunged into the Atlantic late Wednesday night after leaving Kennedy International Airport at 8:17 pm. It carried 215 passengers and 14 crew.
China Airlines, Tabei, China (February 1998)	An A-300 crashed on landing in Taibei, Taiwan and caused 204 fatalities.
Garuda Airlines, Sumatra, Indonesia (September 1997)	An A-300 B4 crashed on approach and caused 234 fatalities.
Korean Air Lines, Guam, USA (August 1997)	A Boeing-747 crashed on landing and caused 227 fatalities.
Collision (Saudia, Kazakstan Airlines), Dadri, India (November 1996)	A Boeing 747-100 (Saudia) was in collision with an IL-76 (Kazakstan Airlines), causing 349 fatalities.
TWA 800, Long Island, New York (July 1996)	A Boeing 747-100 crashed into the Atlantic Ocean off the cost of Long Island shortly after takeoff from Kennedy International Airport. The airplane was on a regularly scheduled flight to Paris, France. The initial reports were that witnesses saw an explosion and then debris descending to the ocean. There were no reports of the flight-crew reporting a problem to air traffic control. The airplane was manufactured in November 1971. It had accumulated about 93,303 flights hours and 16,869 cycles. On board the airplane were 212 passengers and 18 crewmembers. The airplane was destroyed and there were no survivors. Update with design flaw and FAA rewiring regulation.
Birgenair, Puerto Plata, Dominican Republic (February 1996)	A Boeing 757-225 crashed due to instrument malfunction and caused 189 fatalities.
China Airlines, Nagoya, Japan (April 1994)	An A-300-600R crashed during landing and caused 264 fatalities.
Iran Air Tours, Meherabad, Iran (February 1993)	A Tupolev 154 M crashed due to mid-air collision, causing 132 fatalities.
EI AI Boeing 747-200 crash, Amsterdam, The Netherlands (1992)	Separation of both engines from right wing caused the aircraft to plunge fatally into apartment complex, killing the plane's crew and more than 50 people on the ground. Investigations showed that corroded engine mounting bolts led to the catastrophe.
Lauda Air Boeing 767-300 crash, Thailand (1991)	An inadequate thrust reverser deployment flipped a jetliner into a crash dive that killed all 223 people on board.
United Airlines DC-10 crash, Sioux City, Iowa (1989)	The crash due to an engine explosion and the subsequent loss of hydraulic power resulted in the deaths of 111 people on board.

(Continued)

TABLE 2.1 ***(Continued)***
The Fall of Modem Icarus (Major Aircraft Accidents since the 1970s)

Aircraft Accidents	Accident Description
United Airlines Boeing 747 explosion, Hawaii (1989)	Explosive decompression caused by a design that allowed a stray electrical signal to overcome a complex electromechanical system of latches and locks.
Midwest Express Airlines DC-9 crash, Milwaukee, Wisconsin (1985)	An inexperienced and improperly trained crew became disoriented after an engine failed during takeoff and flew the otherwise controllable airplane into the ground.
Japan Airlines Boeing 747 crash, Gumma, Japan (1985)	Crash of a Boeing 747 passenger aircraft caused by metal fatigue from previous repairs found to be faulty and undetected.
Air Canada DC-9 accident (1979)	The rear portion of a DC-9 fuselage ripped open in flight owing to fatigue cracks that had gone undetected in a recent inspection.
American Airlines DC-10 crash, Chicago, Illinois (1979)	Plunged to the ground after left engine fell off and killed 273 people; the engine's rear attachment bulkhead and flanges had been damaged during maintenance procedures. FAA temporarily rescinded the airworthiness certification of the DC-10.
Turkish Airlines DC-10 crash, Ermenonville, France (1974)	All 346 persons aboard a new DC-10 perished when the jet crashed just after takeoff from Paris.

Fractures occur as a result of stress concentrations around imperfections such as cracks, surface roughness due to machining, and crystal misalignments. Many structures are designed for service under high loads. In such cases, the designers must consider the various sources of crack initiation. The mechanisms that lead to stress rupture make up the central subject of fracture mechanics.

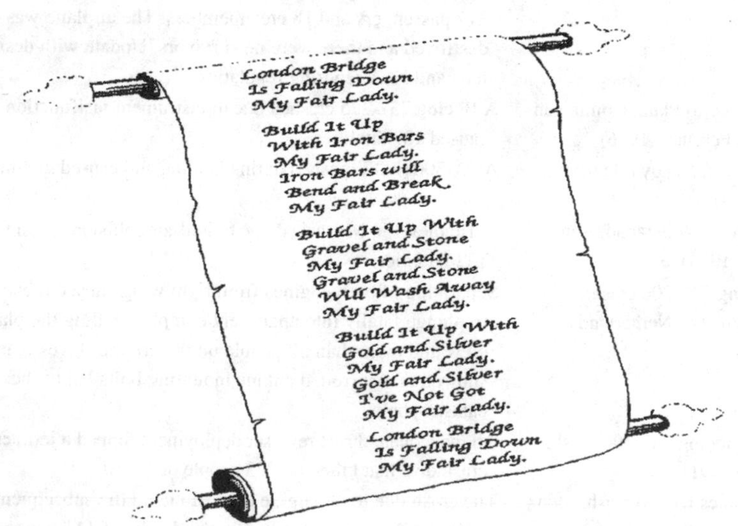

FIGURE 2.2 "London Bridge is Falling Down".

TABLE 2.2
Summary of Overload Failure Modes

Failure Mode	Description	Effects of Failure
Tensile-yield-strength Failure	Occurs under pure tension when the applied stress exceeds the yield strength of the material.	Permanent deformation in the structure.
Ultimate Tensile-Strength Failure	Occurs when the applied stress exceeds the ultimate tensile strength.	Causes total failure of the structure at this cross-sectional point.
Compressive Failure	Similar to the preceding tensile failures only under compressive loads.	Permanent deformation or total compressive failure, causing cracking or rupturing of the material.
Failure due to Shear Loading	Occurs when the shear stress exceeds the shear strength of the material when applying high torsion or shear loads.	Yield and ultimate failures generally occur on a 45° axis with respect to the principal axis.
Brittle Fracture	Certain materials have little capability for plastic flow and are generally brittle, and thus are extremely susceptible to surface flaws and imperfections.	Cracks propagate rapidly and completely through the component when the fracture stress is reached.
Ductile Fracture	High energy needed to continue the fracturing process.	Relatively slow crack growth rates.
Instability Failures	Instability failure occurs in structural members such as beams and columns particularly those made from thin material and where the loading is generally in compression.	A crippling or complete failure of the structure.
Bending Failures	A combined failure where an outer surface is in tension and the other surface is in compression.	Tensile rupture of the outer material.

The Liberty ships, produced in great numbers during the Second World War were the first all-welded ships. A significant number of ships failed by catastrophic fracture. As shown in Figure 2.3, fatigue cracks nucleated at the corners of square hatches and propagated rapidly by brittle fracture. In earlier ships, the riveted plates acted as natural crack arresters. These were absent in the all-welded Liberty ships. The problem was solved by improvements in ship design and steel quality. Investigations revealed that the Liberty ships failed because of three factors: (a) welds were produced by semi-skilled workers, (b) most of the cracks began at square comers, where there was a local stress concentration, and (c) the steel from which the Liberty ships were made had low toughness.

Fracture mechanics has largely evolved from the classic papers of Griffith (1920) and Irwin (1958). Irwin (1983) defines fracture mechanics as: ".... the fracture of materials in terms of laws of applied mechanics and the macroscopic properties of materials. It provides a quantitative treatment, based on stress analysis,

FIGURE 2.3 An example of the fracturing of Liberty ships.

which relates fracture strength to the applied load and structural geometry of a component containing defects". Fracture mechanics provides answers to questions like:

- What is the critical crack size?
- What is the critical energy necessary to propagate a crack?
- How long does it take for a specific crack to grow to a critical crack size?

In fracture mechanics, a stress intensity factor is defined as:

$$K = YS(\pi a)^{\frac{1}{2}} \tag{2.1}$$

where

- S = Nominal stress.
- Y = Geometry factor, which depends upon the shape of the crack, depth of the crack, configuration of the member, and loading conditions.
- a = Crack length.

K, the stress intensity factor, has units of ksi(in)$^{1/2}$ or MN/m$^{1/2}$. Failure occurs when the stress intensity factor exceeds the material fracture toughness,

$$K > K_C \tag{2.2}$$

where K_C is the fracture toughness, a material property. Values are listed here for three common alloys.

$K_C = 25$ ksi $(in)^{1/2}$	7075-T6 Al
80	Ti-6A1-4V
120	4340 steel

The field of fracture mechanics has evolved continuously over the past 40 years. The period from the fifties through the sixties was a time of basic research from a mechanics and materials point of view. In 1968, Rice developed another parameter to characterize nonlinear material behavior ahead of a crack. By modeling plastic deformation as a nonlinear elastic phenomenon, Rice was able to generalize the energy release rate for nonlinear materials. He showed that this nonlinear energy release rate can be expressed as a line integral, which he called the J-integral.

During the seventies, standards and specifications were made. While fracture research in the United States was driven primarily by the nuclear power industry during the 1970s, fracture research in the United Kingdom was motivated by the development of oil resources in the North Sea.

In 1979, a Kurdistan oil tanker broke completely in two while sailing in the North Atlantic. The combination of warm oil in the tanker with cold water in contact with the outer hull produced substantial thermal stress. The fracture initiated from a bilge keel that was improperly welded. The weld failed to penetrate the structural detail, resulting in a severe stress concentration. Although the hull steel had adequate toughness to prevent fracture initiation, it failed to stop the propagating crack.

Fracture toughness can be influenced by temperature in most materials, so that weakening occurs above certain temperatures. Some metals become less brittle as their temperature rises. Two examples are the increase of brittleness of ships in the icy water of the North Atlantic and the increased toughness of nuclear reactor vessels as they heat up to operating temperature from room temperature. Fracture toughness is also lower when loads are applied impulsively, e.g. by impact, as there is no time for plastic deformation to permit local stress relief. The difference can be up to a factor of 2.

Material failure considerations must be broader than just the study of metals as evidenced by the Challenger accident. In January 1986, the Challenger Space Shuttle exploded because an O-ring seal in one of the main boosters did not respond well to cold weather. The Shuttle represents relatively new technology, where service experience is limited. Engineers from the booster manufacturer were concerned about a potential problem with the O-ring seals and recommended that the launch be delayed. Many of us are now aware that rubber O-rings do not function well at low temperature where the rubber loses its resiliency and become quite hard. Unfortunately, these engineers had little data to support their position and were unable to convince their managers or the NASA officials. The tragic results of the decision to launch are well known.

Fracture mechanics theory today is still largely empirical and based upon experimental results. Also, fracture toughness is in practice variable, since we cannot usually know enough about the surface and internal conditions of the material. Therefore, it is not possible to predict fracture strength exactly from the material physics, and allowance must be made for this uncertainty in designing the actual product.

Example 2.1

Semiconductor wafers are extremely brittle. The wafers are cut into small rectangular pieces, each containing an integrated circuit. Each small block of silicon is called a die. In microelectronic manufacturing, the die processing steps include

- Single crystal semiconductor growing;
- Wafer scrubbing;
- Die separating.

A challenging reliability problem is related to the presence of pre-existing micro-cracks in the die, since cracks can form and grow during the above steps.

A vertical crack in the die of an IC package is shown in Figure 2.4. We wish to determine the critical die crack size.

- Nominal stress, S, is, 50.1 MN/m^2;
- Geometry factor, Y, is $2/\pi$;
- Fracture toughness, K_C, is 0.4 $MN/m^{1.5}$.

According to Equation (2.1), the crack size relates to the stress intensity factor by

$$a = \pi^{-1}(K/YS)^{1/2}$$

Thus, the critical die crack size, a_c, can be estimated by

$$a_c = \pi^{-1}(K_C/YS)^{1/2}$$

$$= 5 \cdot 10^{-5}\,\text{m}$$

$$= 0.05\,\text{mm}$$

The critical die crack size is 0.05 mm. Thus, all cracks equal to or larger than that must be screened out during a 100% inspection.

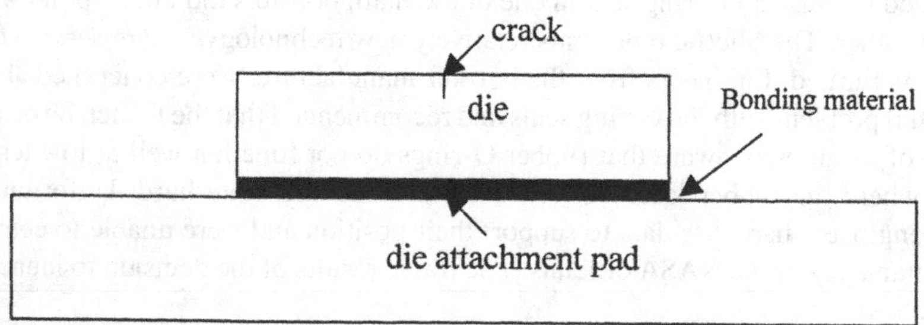

FIGURE 2.4 A vertical crack in the die of an IC package.

2.3 WEAR-OUT FAILURES: CRACK INITIATION AND GROWTH

The Liberty Bell was rung on July 4, 1776 to signal the adoption of the Declaration of Independence. For more than half a century, the Liberty Bell was rung on anniversaries and special events of the new nation until a large crack appeared in 1835 and grew to its now familiar length in 1846. The crack was induced by a flaw produced during casting and some abuse suffered during its transport to be hidden from the advancing British army in 1777.

For wear-out failures, applied loads cause damage that accumulates irreversibly, as in fatigue, wear, corrosion, and dielectric breakdown. The system fails when the damage exceeds the item's endurance level. Accumulative damage does not disappear when the stresses are removed. The main causes of wear-out failures in mechanical components and materials are summarized in Table 2.4.

Fatigue is a frequent cause of wear-out failure. Fatigue failures are caused by cyclical stresses above a critical value and result in more than 80% of all structural failures. The damage is due to internal structural deformation, such as crystal lattice deformation in metal, which does not return to the original condition when the stress is removed. Fatigue damage is cumulative, so that repeated or cyclical stress above the critical stress will eventually result in failure. Just as people get tired, things get worn-out. The science of fatigue of materials deals with one such facet of how things wear out.

The study of fatigue dates backs to the 19th century. At that time the growing railroad industry was being plagued by axle failures, despite the fact that loads never exceeded the engineer's predicted maximum load – the yield point. So the idea arose that a structure may fail due to cyclic loading (loading-unloading sequences) even if these loads never exceed the static material strength, i.e., yield point. This idea is the basis for the study of fatigue of structures.

Like the trains of the 19th century, turbines and other high-speed machinery are subject to continuous loading and unloading. Take a wind turbine as an example; the rotation of the rotor, shafts, and gears, the gusting and lulls of the wind, and subsequent vibrations induced by all these factors and more lead to load cycles varying in size, frequency, and sequence. These repetitive and rapidly changing loads can excite the natural resonance of the structure and, if the turbine is improperly designed, can lead to catastrophic failure – the turbine literally can shake itself apart. Even under ordinary operating conditions, the cyclic nature of these loads can limit the design life time of the structure.

The fatigue process consists of the following two stages:

- *Crack Initiation*: Microscopic cracks develop at "nucleation sites" - points of material weakness or stress concentration. As repeated loading continues, these cracks grow and some coalesce into a dominant macroscopic fatigue crack.
- *Crack Propagation*: The fatigue crack grows at an accelerating rate as the loading cycles continue. If the crack reaches an intolerable size for the load being applied, the weakened structure can no longer withstand the stress and a fatigue failure occurs.

TABLE 2.3
Data from *S-N* Fatigue Tests

S_i, Stress (ksi)	N_i, Fatigue Cycles
60.000	20,000
50.000	30,000
40.000	25,000
40.000	60,000
25.000	70,000
30.000	200,000
20.000	300,000
25.000	700,000
15.000	400,000

To determine the fatigue properties of a material, tests were devised where a specimen is cyclically loaded to a given stress, *S*, until it fails (Table 2.3). Then counting the number of load cycles to failure, *N*, it is determined how many cycles of load size of *S* will cause failure of the material. Repeating this test for numerous loads and plotting them against the number of cycles to failure yield a material

TABLE 2.4
Summary of Wear-out Failure Modes

Failure Mode	Description	Effects of Failure
Fatigue Failure	Repeated loading and unloading of a component, with cycling loads regarding the maximum load and number of cycles are the predominant variables.	Reducing the design allowable stress below those for static properties. Limiting design life in many applications.
Metallurgical Failure	Caused by extreme oxidation or operation in corrosive environments.	Material/structural failures.
Mechanical Wear	Failures occur when surfaces moving in contact are damaged.	Higher friction and further damage.
Failure due to Stress Concentration	Failures occur due to an uneven stress, while stress concentrations take place at abrupt transitions from thick gauges to thin gauges, at abrupt changes in loading along a structure, at right-angle joints, or at various attachment conditions.	Failures at stress-concentration locations.
Failure due to Flaws in Materials	Failures due to improper inspection of materials, weld defects, fatigue cracks, and other flaws.	Reduce the observed strength of material and result in premature failure at the flawed location.
Creep	Permanent elongation due to the formation of a network of microcracks, due to combined tensile stress and temperature effects.	Reduced strength, as well as dimensional changes and deformation in metals.

fatigue property chart, known to fatigue gurus as the *S-N* curve. As the stress is reduced, the number of cycles to failure increases. The *S-N* curve can be used to estimate a product's life time under a given loading condition, as illustrated by the following example.

Example 2.2

From the fatigue-test data given below, we will construct the *S-N* curve for a turbine part. Assuming the part completes 10,000 fatigue cycles per year at an operating stress of 18 ksi, estimate the product's useful service life.

These data tend to fall along a straight line if we plot log$S$ vs. logN_C. Fatigue strength model has the following general form:

$$NS^m = A \tag{2.3}$$

For the data shown in Figure 2.5, values of m and A can be obtained from a least square analysis which yields:

$$m = 2.596$$

$$A = 7.72E8$$

For the operating load of 18 ksi, the fatigue cycles to failure are calculated to be

$$N = A/S^m = 7.72E8/(18)^{2.596} = 4.26E5$$

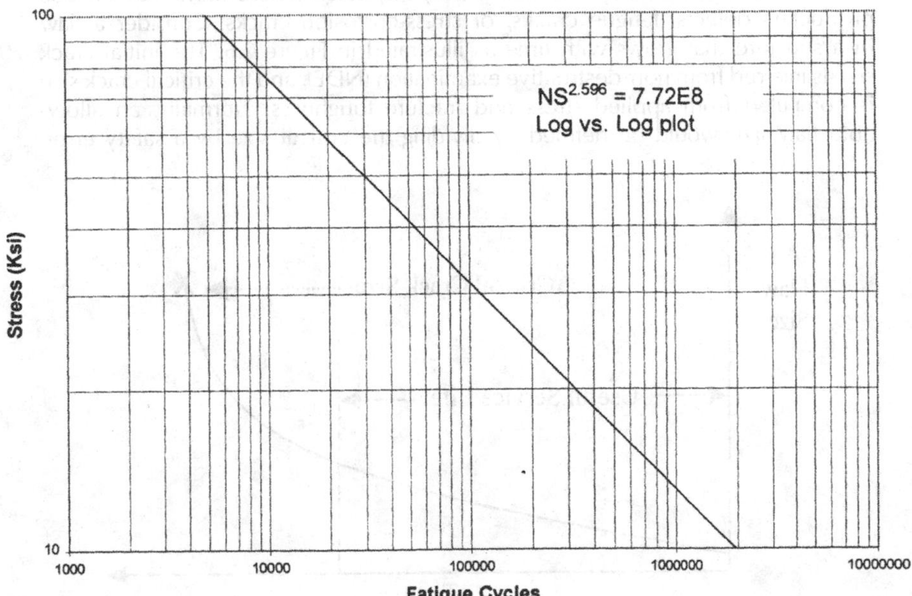

FIGURE 2.5 *S-N* curve: Determine fatigue life according to load conditions.

Since the turbine part completes 10,000 cycles per year, the product useful service life is

$$L = 4,26E5/1E4 = 42.6 \text{ years}$$

This example demonstrates how to estimate a product's useful life, given the operating stress. However, the random nature of the fatigue loading poses a unique challenge to the engineer. For a wind turbine, how can we know how long it will last if we cannot know precisely the loads it will be subjected to. Furthermore, to increase the life of a wind turbine, engineers need to know what is causing them to break in the first place. Is it the incessant small vibrations that accompany normal use, or the rare but large loads of gale force winds? What will help more: stronger, more expensive materials or computerized control systems monitoring the machine's behavior?

The answer lies in combining statistics and engineering. While we cannot know precisely what the wind strength will be at any point of time, we can certainly narrow down the possibilities. With knowledge of the statistical distribution of wind speed, a statistical distribution of turbine loads can be predicted. Then using an *S-N* diagram, a fatigue life can be estimated, from which a statistical probability of attaining such longevity can be determined.

Fatigue failure can thus be theoretically avoided, but the above approach often leads to over-designing structures. Such massiveness would be totally inappropriate in aircraft design where the scales must be carefully balanced between structural safety and payload capability. The mandatory requirement for the routine inspection of aircraft engines and structures furnishes a necessary process for the detection, control, and rectification of incipient signs of corrosion and/or cracking.

Damage tolerance, as its name suggests, entails allowing sub-critical flaws to remain in a structure that is periodically inspected. These flaws may include production defects, fatigue cracks, or stress-corrosion cracks. Consider a flaw in a structure that grows with time as illustrated in Figure 2.6. The initial crack size is inferred from non-destructive examination (NDE), and the critical crack size is computed from applied stress and fracture toughness. Normally, an allowable flaw size would be defined by dividing the critical size by a safety error.

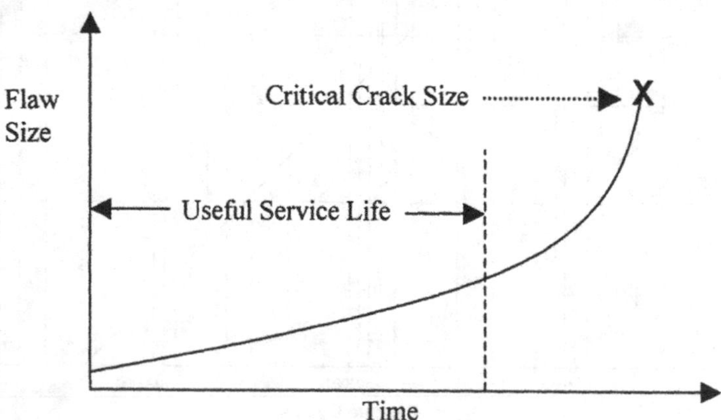

FIGURE 2.6 The damage tolerance approach to design.

The allowable flaw size should be such that it will not grow to the critical size during the period between inspections. The predicted service life of the structure can then be inferred by calculating the time required for the flaw to grow from its initial size to the maximum allowable size.

The fatigue of an aircraft structure can be mathematically modeled based on linear elastic fracture mechanics theory. The maximum size of undetected small flaws can be hypothesized from the observed strength of the structure or inferred from the sensitivity of the testing techniques. This model allows for the variability of crack initiation and crack growth found in the materials used for aircraft structures. Combinations of material crack growth parameters are selected from a distribution by Monte Carlo simulation and then used to describe the stochastic behavior of crack growth.

A similar approach for structural strength degradation has been extended to multiple failure modes, including corrosion growth in aircraft structures. This extension is based on the fact that aircraft structures can be conveniently broken down into two structural element types: one representing a fuselage frame and the other representing a wing stringer. The probabilistic approach coupled with in-service operating experience allows prediction of the probability of finding defects of different magnitudes during standard inspections. Inspection routines are thus established to provide suitable values of structural reliability during the operational life of a fleet of airplanes.

In the case of turbine aircraft engines, the turbine blading is critical. Thermal fatigue, creep, and fatigue from high-frequency vibration are the predominant failure modes. Although laboratory simulations can give some confidence to the design during pre-service testing, the best data obtainable are those gained from in-service experience and related information feedback. Here, the failure type of greatest concern is creep and creep rupture. Creep damage accumulation (and crack initiation) can be evaluated by laboratory examination of blading removed from the engine during routine inspections. Total blade sets are replaced when evidence based on the above process indicates the need.

2.4 ENVIRONMENTAL IMPACT: TEMPERATURE-RELATED FAILURE

Prolonged use at elevated operating temperatures can lead to failures in many objects. One particularly temperature-sensitive group of objects is that of electronic components where failures are due to creep in the bonding materials, parasitic chemical reactions in switches and connectors, and diffusion in solid-state devices. Electrical failures may also be caused by electrical overstress, dielectric breakdown, and electromigration.

Arrhenius studied reaction rates as a function of temperature, and in 1889, he introduced the concept of activation energy as a measure of the temperature sensitivity of the reaction rate for specific chemicals. Chemical reactions proceed at different speeds at different temperatures. The reaction

$$\text{Gasoline} + \text{Oxygen} \rightarrow \text{Carbon Dioxide} + \text{Water}$$

proceeds very slowly at room temperature. That's why you can pour gasoline into your car without it bursting into flames. At high temperatures, it happens quickly. The reaction also releases a large amount of heat. That's why, if you light a match while pouring gasoline into your car, it can cause a problem. The gasoline nearest your match heats up, reacts with the oxygen in the air, and gives off enough additional heat to raise the temperature of the next bit of gasoline. That's why they have NO SMOKING signs at gas stations.

For first-order (simple) chemical reactions, the temperature dependence of reaction rate is of the Arrhenius form:

$$\text{Reaction Rate} = \text{C} \exp(-E_a/kT).$$

where T is the temperature on an absolute scale, k is Boltzmann's constant, and E_a is the activation energy of the chemical reaction. This exponential function can represent very large variations of rate with temperature changes. Even on a hot day, gasoline vapors in an enclosed space still smell after many hours (DON'T TRY THIS AT HOME), so we know the reaction rate is very slow and E_a is much bigger than kT even for hot days.

Many physical and chemical processes have a temperature dependence similar to the Arrhenius expression with the activation energy characterizing the sensitivity to temperature. Because failure mechanisms in a device are basically physical or chemical degradation processes, the dependence of a failure mechanism on temperature can often be modeled by the Arrhenius model. Knowing that failure processes are accelerated by raising temperature, one can shorten the test time for an object by testing it at an increased temperature. If we assume that the process that ultimately leads to failure has its rate given by:

$$\text{Rate} = \text{A} \exp(-E_a/kT)$$

and that the time to failure is inversely related to process rate:

$$T_f = \frac{C}{Rate} = B \cdot exp\left(\frac{E_a}{kT}\right)$$

If a component fails at time at test temperature T_1 and at time t_2 at operational temperature T_2 ($T_1 > T_2$), then

$$t_2/t_1 = exp\left[E_A/k\left(1/T_2 - 1/T_1\right)\right] \tag{2.4}$$

where t_2/t_1 is the acceleration factor. According to the Arrhenius model, the higher the test temperature, the shorter the test time required to demonstrate a certain design life at operational temperature.

Example 2.3

For a microprocessor, the test temperature is T_1 = 125 + 273.16 = 398.16°K and the design temperature is T_2 = 55 + 273.16 = 328.16°K. For a specific type of failure mode, the activation energy is found to be E_A = 1.0 eV. A sample unit survived t_1 = 30 hours under test temperature T_1. Estimate the design life under operational temperature T_2.

According to Equation (2.5),

$$t_2/t_1 = \exp\left[E_A/k(1/T_2 - 1/T_1)\right] = \exp\left[1.0/8.671E-5(1/328.16 - 1/398.16)\right]$$
$$= 501$$

where $t_1 = 30$ hours. So

$$t_2 = 501 \times 30 = 15 \times 10^3 \text{ hours}$$

From the accelerated testing, the design life is estimated to be 15,000 hours.

The Arrhenius rate law is usually used in its logarithmic form:

$$\ln t_f = \ln A - (E_A/R)(1/T)$$

A plot of $\ln t_f$ against $1/T$ gives a straight line whose slope is E_A/R and whose intercept is $\ln A$. The Arrhenius model can be used if only one single failure mechanism is present, so that E_a has a constant value.

The Arrhenius model is widely used to model product life as a function of temperature. Applications include

- Electrical insulation and dielectrics;
- Solid-state and semiconductor devices;
- Battery cells;
- Lubricants and greases;
- Plastics;
- Incandescent lamp filaments.

Creep is one of the many types of material behavior that obeys an Arrhenius relationship. As shown in Figure 2.7, a plot of the creep rate (on a log scale) vs. $1/T$ where T is the absolute temperature produces a straight line. One practical use of this is in performing accelerated creep testing. Measuring the creep rate at two

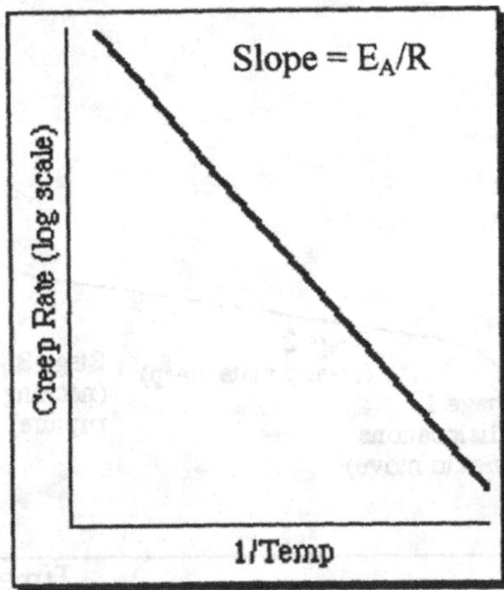

FIGURE 2.7 Arrhenius model of creep.

elevated temperatures allows extrapolating to the expected rate at a service temperature, e.g., to predict the life of parts such as turbine blades. The slope of the Arrhenius curve can provide the activation energy for the creep process.

Creep may be defined as a time-dependent deformation at elevated temperature and constant stress. At elevated temperatures (more than about one-half the melting point), materials will deform plastically at applied stresses below the yield stress determined with a low-temperature tensile test. A typical creep test is performed by applying a constant load and measuring the strain (elongation) as a function of time. The resulting curve shows three stages. During the first stage, dislocations climb and break free from whatever was pinning them. The second stage of creep is characterized by movement of dislocations resulting in a steady rate of strain. In the third stage, necking and failure occur. The creep rate in the second stage of creep is an Arrhenius function of temperature. The creep rate is higher at a higher temperature and so failure happens sooner. At a higher stress, the initial strain is higher, which also reduces the time for failure to occur.

The end of useful service life of the high-temperature components in a boiler (the super-heater and re-heater tubes and headers, for example) is usually a failure by a creep or stress-rupture mechanism. The root cause may not be just elevated temperature, as fuel-ash corrosion or erosion may reduce the wall thickness so that the onset of creep and creep failures occurs sooner than expected.

However, regardless of the cause, the failure will exhibit the characteristics of a creep or stress rupture. The *ASME Boiler and Pressure Vessel Code* recognizes creep and creep deformation as high-temperature design limitations and provides allowable stresses for all alloys used in the creep range. One of the criteria used in the determination of these allowable stresses is 1% creep expansion, or deformation, in 100,000 hours of service. Thus, the code recognizes that over the operating life, some creep deformation is likely. And creep failures do display some deformation or tube swelling in the immediate region of the rupture. The three stages of creep are illustrated in Figure 2.8.

The temperature at which creep begins depends on the alloy composition. For the common materials used in boiler construction, Table 2.5 gives the approximate temperatures for the onset of creep. It should be pointed out that

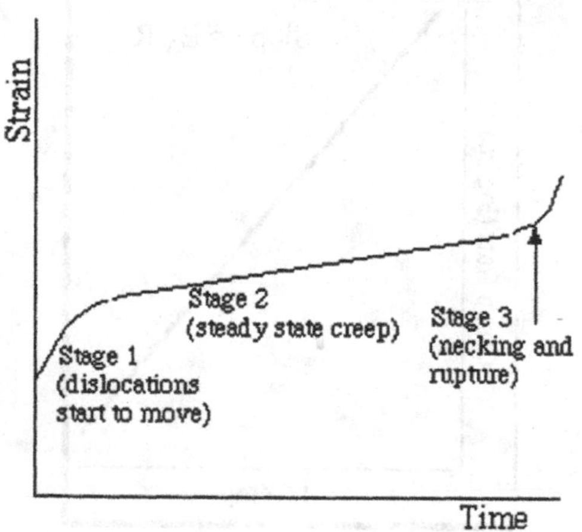

FIGURE 2.8 Three stages of creep.

TABLE 2.5
Initial Creep Temperature

Carbon steel	800°F
Carbon +1/2 molybdenum	850°F
1-1/4 chromium-1/2 molybdenum	950°F
2-1/4 chromium-1 molybdenum	1,000°F
stainless steel	1,050°F

the actual operating stress will, in part, dictate or determine the temperature at which creep begins.

Temperature influences product reliability in many different ways. When steel is cooled below the reference temperature, brittle fracture can occur. This was partly responsible for the sudden collapse during the winter of some early welded bridges and the spectacular breakup of Liberty ships during Second World War. The ductile-brittle transformation because of the cold weather was also a contributor to sinking of the Titanic, which impacted a glacier during its maiden voyage. For electronic components, many failure mechanisms are functions of temperature change, the rate of temperature change, and spatial gradients as well.

2.5 SOFTWARE AND RELATED "HARD" FAILURES

Software is increasingly being introduced into safety-critical systems and, as a consequence, has been involved in accidents. Software failure due to undetected errors in the program can cause catastrophic consequences. Let's take a look at a classic, oft-cited, example of an accident that involves software and human errors, that of the Therac-25. The Therac-25 was a million-dollar radiation machine designed to precisely aim a beam of radiation at a patient in order to treat tumors or cancerous growths. Patients were often recovering from operations that had removed the bulk of a tumor, and underwent these radiation treatments to kill tumor cells left behind.

The Therac-25, designed by Atomic Energy of Canada Limited (AECL), was a high-energy radiation machine, but radiation treatment usually involved many low-energy dosages across successive treatment sessions. The machine was controlled through a computer located in another room (as are most radiation therapy controls, in order to protect the technicians from unnecessary exposure).

There were two basic modes in which the Therac-25 could function (see Figure 2.9). The first was the low-energy mode mentioned above, in which an Electron beam of about 200 rads was aimed at the patient and sent off in a short burst. The second mode was an X-ray mode generating more penetrating radiation for more deeply located tumors. This mode used the full 25 million electron volt capacity of the machine. When the machine was switched into this mode, a thick metal plate would be inserted between the beam source and the patient; as the beam struck the plate, it was transformed into X-rays that would radiate tumors and the like.

To switch to "electron mode", the technician typed "**e**" at the computer terminal. To switch to "x-ray mode", the technician typed "**x**" at the computer terminal.

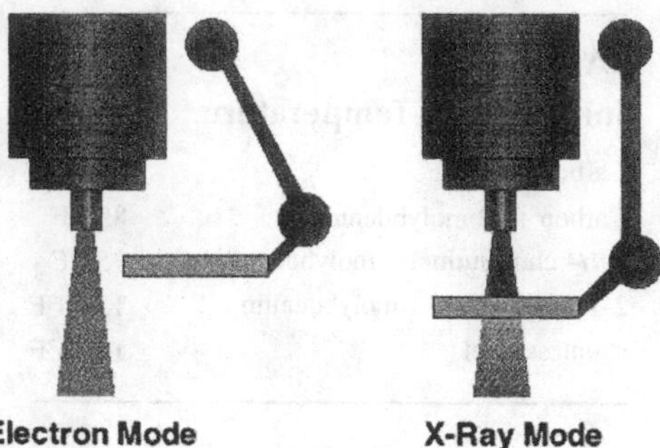

FIGURE 2.9 Two basic modes of a radiation machine.

Occasionally, the Therac operator mistyped an "x" for an "e", but noticed the error before triggering the beam. An "edit" of the input data was performed by using the "arrow up" key to move the cursor to the incorrect entry, changing it, and then returning to the bottom of the screen, where a "beam ready" message was the operator's signal to enter an instruction to proceed, administering the radiation dose.

In 1986, a Texas oil worker went in for his usual radiation treatment for a tumor he had removed from his left shoulder. He had been here eight times before, so this was business as usual. They got him set up on the table, and the technician went down the hall to start the treatment. The technician sat down at the terminal and hit "x" to start the process. She immediately realized that she made a mistake since she needed to treat the patient with the Electron beam, not the X-ray beam. She hit the "Up" arrow, selected the "Edit" command, hit "e" for Electron beam, and hit "Enter", signifying she was done configuring the system and was ready to start treatment.

In a real-time system like the Therac-25, the software must operate at the speed demanded by the system inputs and outputs. The total time for the interaction described above was under 8 seconds. It turns out that this particular sequence of actions within this timeframe had never occurred in all of the testing and evaluation of the Therac-25. If it had occurred, it would have pointed out a dangerous bug in the system. Since most programs consist of many individual statements and logical paths, the scope for errors is large. If errors exist in a portion of the software that is frequently executed, the errors will be easy to find during testing and the probability of causing failure after software release will be low. However, if errors exist in a part of the software rarely executed by its users, the errors will be very difficult to discover by testing and the probability of causing failure will be extremely high.

In this particular case, the system presented the technician with a "Beam Ready" prompt, indicating it was ready to proceed; she hit "b" to turn the beam therapy on. She was surprised when the system gave her an error message since she had never run into this scenario before.

Software failures are often linked with human errors. In this case, the technician wasn't familiar with this particular error message, but these particular errors usually meant that the treatment hadn't proceeded. She cleared the error to reset the Therac-25 so she could do it again. She got the "Beam Ready" prompt and again hit "b" to initiate the treatment. Same deal: an error message and the system stopped. She tried it again.

Software can cause hard failures! Meanwhile, back in the treatment room, the patient was feeling repeated burning, stabbing pains on his back. None of the previous treatments had been like this. However, the operator was isolated from the patient since the machine apparatus was inside a shielded room of its own. The only way the operator could be alerted to patient difficulty was through audio and video monitors. On this day, the video display was unplugged and the audio monitor was broken. Although he cried out several times, asking (first jokingly) whether the system was configured right, no one came to check on him. Finally, after the third painful burst, he pulled himself off the table and went to the nurses station.

In an operational system such as a process controller or an auto-pilot, it is essential that the software is ready to accept inputs and completes tasks at the right time. The problem here with the Therac-25 was this: when the particular sequence of commands was executed quickly enough (e.g., in under 8 seconds), the arm correctly withdrew as it should in the Electron beam mode, but the beam switch never occurred. Although the machine told the operator it was in the Electron beam mode, it was actually in a hybrid proton beam mode. As a result, the system was delivering a radiation blast of 25,000 rads with 25 million electron volts, more than 125 times the normal dose. As shown schematically in Figure 2.10, the particular sequence of steps executed by the technician had moved the metal plate from the beam's path, but left the power setting on maximum.

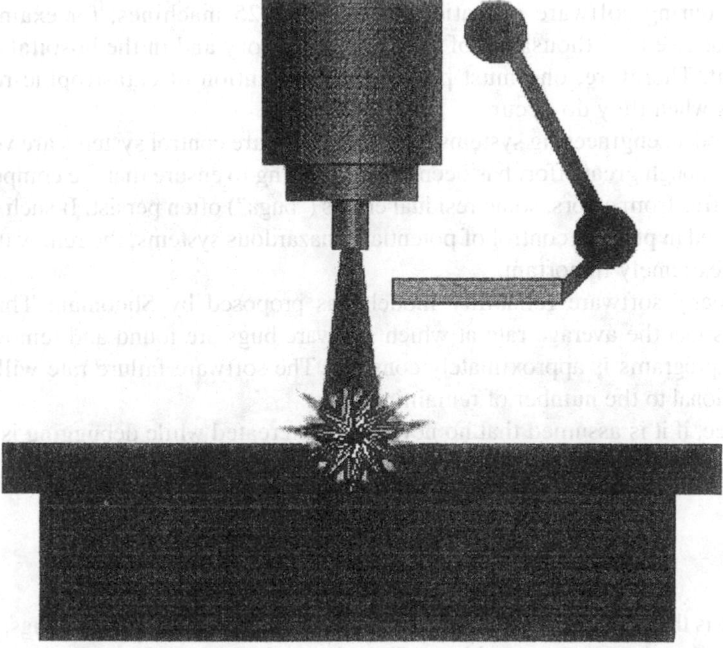

FIGURE 2.10 Overdoses: Failure is never soft!

Software risk depends not only upon the existence of errors but also upon the probability that an existing error will affect the output. In this case, the patient's health deteriorated rapidly from radiation bums and other complications from the treatment overdose. He died four months later. It is worth noting that the problem wasn't actually diagnosed until three weeks later, when it happened again to another patient. At this point, the senior technician realized something about the sequence of steps being taken that must be triggering this flaw. After investigation, he found the problem with the plate and reported it to the manufacturer. Subsequent investigation showed that there had been similar overdoses in Georgia, Washington, and Canada.

When a software programming error does exist, it exists in all copies of the program. Furthermore, if it is such as to cause failure in certain circumstances, it will always fail when those circumstances occur. Therefore, software errors can be extremely serious. Between June 1985 and January 1987, six known accidents involved massive overdoses by the Therac-25 – with resultant deaths and serious injuries. They have been described as the worst series of radiation accidents in the 35-year history of medical accelerators.

As summarized in Table 2.6, software errors ("bugs") can arise from the specification, the software system design and from the coding process. The Therac-25 incident serves as a good example of what are called latent errors with software systems. These are "accidents waiting to happen", errors that are virtually offered by the software system. For complex interrupt-driven software, timing is of critical importance. It is unlikely that software testing will discover all possible errors that involve operator intervention at precise timeframes during software operation. The Therac-25 machines, for example, had been exercised for thousands of hours in the factory and in the hospitals without accident. Therefore, one must provide for prevention of catastrophic results of failures when they do occur.

In modem engineering systems, real-time software control systems are very common. Although great effort has been spent on trying to ensure that the computer software is free from errors, some residual errors ("bugs") often persist. If such software is involved in process control of potentially hazardous systems, the removal of these bugs is extremely important.

An early software reliability model was proposed by Shooman. This model assumes that the average rate at which software bugs are found and removed from similar programs is approximately constant. The software failure rate will then be proportional to the number of remaining bugs.

Hence, if it is assumed that no new bugs are created while debugging is in progress and that all detected errors are corrected, we may write

$$\varepsilon_T = \left(\frac{E_T}{I_T}\right) - \varepsilon_C(\tau) \qquad (2.5)$$

where τ is the debugging time; ε_T is the fractional number of residual bugs; ε_C is the fractional number of corrected bugs; E_T is the total number of initial errors; and I_T is the total number of instructions.

TABLE 2.6
Summary of Software Failure Modes

Phase of Software Development	Software Failure Modes
Specification	Incorrect interpretation of the requirement;
	Logically incomplete – not cover all the possible input conditions and output requirements;
	Inconsistent – give conflicting information or use different conventions in different sections;
	Cannot be validated (e.g., include requirements that are not testable).
Design	Incorrect interpretation of the specification;
	Incomplete or incorrect logic.
Coding	Typographical errors;
	Incorrect numerical values, e.g., 0.1 for 0.01;
	Omission of symbols, e.g., parentheses;
	Inclusion of expressions which can become indeterminate, such as division by a value which becomes zero;
	A compiler, test package, or version-checking tool has defects, which causes the code created to be defective.
System Integration/ Configuration Control	Software modules are omitted when the product's components are integrated.
	The passing of data between modules or the accessing of external tables is incorrect.
	The wrong versions of modules are used.
Operation	An incorrect edit procedure in one program allows faulty data to be used in another program.
	A table of values is searched using a faulty index; thus, an incorrect location is updated.
	A marginal item of hardware changes one or more bits in storage and thus changes the software during operation.
	Radiation-produced events (due to X-rays, radioactivity, or cosmic rays) change one or more bits in storage.

Let us further assume that the failure rate λ, after a period of debugging τ, will thereafter remain constant and proportional to the number of residual errors $\varepsilon_T(\tau)$:

$$\lambda = K\varepsilon_T(\tau)$$

where K is a constant of proportionality. The failure probability for the program during the program operating time interval Δt may be written as

$$F(\Delta t) = 1 - e^{-\lambda\Delta t} = \lambda \cdot \Delta t = K\varepsilon_T(\tau)\Delta t \tag{2.6}$$

Equations (2.5) and (2.6) are thus the basis of Shooman's model. His experimental findings suggest that the value for (E_T/I_T) is approximately constant and lies in the range 1.0–3.0*E*-2. The fractional number of corrected bugs ε_c may be described by

$$\varepsilon_c(\tau) = \rho\tau$$

where *p* is the fractional rate at which errors are removed and the debugging time τ is in person months. A value for p in the range 1.0–3.0*E*-3 per month is suggested through Shooman's experimental work.

Shooman's model represented an early attempt to quantify the software risk. However, this approach has its limitations; the constant of proportionality K must be calculated for each individual program by functional testing of the program since K is dependent on the dynamic structure of the program and the degree to which faults are data-dependent. Musa and other researchers have proposed many new models for software reliability.

A software safety and risk reduction approach must be technically and economically sound to be embraced by program management and engineering disciplines. It must also meet the intent and spirit of existing software and safety criteria and standards. The following risk engineering approach maximizes safety risk control and increases quality assurance.

2.5.1 Identification of Hazards and Software Causal Factors

System and subsystem hazards and failure modes are identified which are caused by direct software inputs or software-influenced human error. This also includes functional, physical, zonal, and environmental interface hazards and failure modes.

2.5.2 Identification of Software Safety Requirements

System hazards and failure modes influenced by software are communicated to systems and software engineering. Safety-specific requirements are derived and implemented into preliminary and detailed design.

2.5.3 Identification of Testing and IV&V Requirements

Specific testing and IV&V requirements are identified for the purpose of verifying the elimination or control of software-influenced hazards and failure modes. The testing must verify that safety requirements were implemented in the design and control safety risk to acceptable levels.

The role of the engineer is to respond to a need by building or creating something along a certain set of guidelines (or specifications) which performs a given function. Just as importantly, that device, plan, or creation should perform its function without fail. Everything, however, most likely will eventually fail (in some way) to perform its given function with a sought after level of performance. Hence, the engineer must struggle to design in such a way as to avoid failure, and, more importantly, catastrophic failure. The engineer must design out failures that could result in loss of property, damage to the environment of the user of that technology, and possibly injury or loss of life. Through analysis and study of engineering failures and their mechanisms, modem engineering designers can learn what to avoid and how to create designs with less chance of failure. Well-designed systems will have very reliable components, will be fault-tolerant (system still works when components fail), and will be "fail safe" (system will go to a "safe" state when it does fail).

2.6 ARTIFICIAL INTELLIGENCE (AI) AND ITS SHOCKING FAILURES

It should be recognized that the topics of machine learning (ML) and artificial intelligence (AI) are not just based on computer science; they also draw on computer engineering and a number of other disciplines as evidenced by the Reference. While the majority of computer science research focuses on solving software-related problems, such as how to use several programming languages, operating systems, and databases, the main goals of computer engineering are solving real-world engineering problems, including hardware and software interfaces.

As a proof, quantum computing, quantum information, and quantum communication, which integrate new hardware and software infrastructures, are making great impacts to ML and AI. For example, D-Wave Systems quantum computers have control features that allow users to tune the quantum computational process and solve problems faster with more diverse solutions. The business unveiled its most sophisticated quantum computing system on September 27, 2016, with a 2000-qubit processor that doubles the speed of its predecessor.

The broad perspective/mindset about ML and AI, necessary to the success of every engineering practitioner including engineers and engineering managers, is addressed here. The following case studies and examples provide an excellent venue to fulfill the need since it is engineering bringing ML and AL to real world for practical applications for the benefits of the people.

2.6.1 Case Study: A320 Warsaw Crash

For AI/ML, it is important to recognize the risk of dangerously assuming "the issues are known in the field". Therefore, this section discusses "A320 Warsaw Crash" as a Case Study. For this fatal accident, each mechanical/electrical/electronic component was operated without failure or deviation; it's thus classified as one of the "non-failure accidents". The catastrophic accident was caused by, after touchdown, the software-based model which incorrectly believes that the aircraft has not landed due to insufficiency with algorithms failing to consider the off-nominal landing conditions at Warsaw resulting in plane overruns, crashes, and catches fire.

The section helps professionals in industries and related academic fields to fill the gap in their knowledge about "non-failure accidents", understanding software's "Fatal" insufficiency causing catastrophic accidents.

An Airbus A320 made a stormy landing at Warsaw Airport in Poland on September 14, 1993. It went over the runway's terminus, climbed a bank of soil, and finally came to rest there. In this catastrophe, there were two fatalities and several injuries.

2.6.1.1 Description of the Software of the Aircraft System

The software must make sure the aircraft is on the ground even if the systems are selected mid-air in order to guarantee that the thrust-reverse system and the spoilers are only active in a landing situation. Automation algorithm allows thrust reversers when

- At least 6.3 tons on each main landing gear strut

Automation algorithm allows ground spoilers when

- At least 6.3 tons on each main landing gear strut

or

- Wheel turning at least 72 knots

Automation algorithm triggers wheel brake when

- Wheel turning at least 0.8 V0 knots

Only when the weight of each main landing gear strut is at least 6.3 tons or when the plane is traveling at a speed greater than 72 knots (133 km/h; 83 mph) are the spoilers engaged.

Only if the first requirement (when the weight of each main landing gear strut is at least 6.3 tons) is satisfied are the thrust reversers turned on. The pilots cannot manually engage either system in order to override the software's choice.

The most effective braking mechanism was not engaged in the Warsaw accident because none of the first two requirements were met. The required pressure of 12 combined tons on both landing gears needed to activate the sensor was not met since the plane landed at an angle (to combat the anticipated crosswind). Due to hydroplaning on the slick runway, the plane's wheels were unable to rotate at the required minimum speed.

2.6.1.2 Causes of the Accident and Insufficiency with Software Algorithms

The catastrophic accident was caused by, after touchdown, the software-based model which incorrectly believes the aircraft has not landed due to insufficiency with algorithms failing to consider the off-nominal landing conditions at Warsaw resulting in plane overruns, crashes, and catches fire.

The autonomous aircraft systems didn't enable the ground spoilers and engine thrust reversers to work until the left landing gear made contact with the runway. The airplane was unable to stop before the end of the runway due to the braking distances in the severe rain. The plane touched down, but the computer didn't actually acknowledge it until 125 meters past the halfway mark of the runway.

2.6.1.3 Consequence

One of the passengers was killed when a fire started inside the cabin as a result of the impact. The incident also claimed the life of the co-pilot. Five people were slightly hurt and 51 people were gravely hurt, including two crew members.

2.6.2 Examples: Why Artificial Intelligence Fails?

2.6.2.1 AI to Treat Cancer May Result in Patient Death

IBM lost US$62 million on yet another failure in trying to create an AI system that would help fight cancer. But the result was once more disappointing. A physician at Jupiter Hospital in Florida claimed that the product was a total failure. He continued by saying that they bought it for marketing reasons. According to medical professionals and consumers, Watson encouraged doctors to give a cancer patient who was experiencing severe bleeding a drug that would make the bleeding worse. Medical professionals and patients both reported numerous instances of harmful and incorrect treatment recommendations.

2.6.2.2 A Mask can Fool AI for Secure System Access by a Face

Facial recognition is cropping up everywhere nowadays, but it may not be as secure as we initially thought. Researchers have been able to find instances in which facial recognition has been fooled using a 3D-printed mask that depicts the face used to authenticate the Facial ID system.

If you have an iPhone X with Face ID, make sure no one is wearing a mask over your face. Apple claims that Face ID uses the iPhone X's potent front-facing camera and ML to build a 3D representation of your face. The ML/AI component enabled the system to maintain security while adapting to cosmetic changes (such as donning make-up, putting on glasses, or wrapping a scarf around your neck). A security company in Vietnam named Bkav revealed how to effectively unlock an iPhone with Face ID by adding 2D "eyes" on a 3D mask. The stone powder mask was made, and it cost about $200. Simple infrared images printed on paper served as eyeballs.

Wired, however, tried to use masks to bypass Face ID but was unable to duplicate the findings.

2.6.2.3 Death from an Uber Self-Driving Car

Elaine Herzberg, who was the first pedestrian fatality in a self-driving car, will be remembered on March 18, 2018, as a day in history to honor her life.

It happened in Tempe, Arizona, in the United States. Herzberg was fatally injured while attempting to cross a four-lane highway on a bicycle. Because Uber was aware of the risks associated with autonomous vehicles, they built a human-in-the-loop system as a safety net. According to reports, the safety driver, however, may have missed up to a third of the trip because she was preoccupied with a voicemail on her phone.

The AI was unable to identify an object as a pedestrian until it was close to a crosswalk, which is what led to the event.

The investigation revealed, among other things, that Uber's internal safety risk assessment processes and operator monitoring were deficient, and that turning off the vehicle's automatic emergency braking and forward collision warning systems increased risks.

Uber ceased the testing of self-driving cars in Arizona after the tragedy, which was tragic and understandable, given the location where such testing had been authorized since 2016.

2.6.2.4 Mars Polar Lander

Software glitch caused the spacecraft to collide with Mars at a lower height than expected, whereupon it was destroyed by atmospheric stress.

The legs were extended at a height of 40 meters during the fall to Mars. A brief signal was sent by the touchdown sensors (on the legs). The required response from the software was to stop the descending engines. At 50 mph, the vehicle free-fell and destroyed upon impacting the ground.

When the lander's legs made contact with the ground, micro-switches built into the legs were supposed to notify the onboard engines to shut off. However, as the landing legs spread out during the fall through the Martian atmosphere, the switches felt a jolt, and malfunctioning software was unable to distinguish between the jolt and an actual touchdown. Because there were no retro-rockets to cushion the impact, the engines cut off too soon, causing the lander to fall the remaining distance to the surface.

This example demonstrates the significant effects that poor requirements, specifications, and guidelines may have on software development. In this instance, NASA was already in possession of a Software Interface Specification (SIS), which outlines the precise format and measurement units for the software utilized by ground-based computers. The calculations for the Mars Climate Orbiter were incorrect by a factor of 4.45, which is the conversion factor from force in pounds to Newtons, since the contractor (Lockheed Martin) did not adhere to the SIS. This inaccurate data was used to construct an improper trajectory.

What a setback for space exploration to spend $327.6 million on this mission only to have the probe crash into the Martian atmosphere because two software engineering teams, who both produced excellent code, failed to cooperate and adhere to the project's requirements.

2.6.2.5 Boeing 787 Lithium Battery Fires

Reliability analysis predicted 10 million flight hours between battery failures. Within 52,000 flying hours, however, two fires were caused by battery failures. Three other, less-reported instances of smoke in the battery compartment are not included in this.

It all began in January 2013, when a Japan Airlines 787 that was empty and parked at Boston's Logan Airport caught fire. The entire fleet of Dreamliners was grounded by the Federal Aviation Administration, while the issue was repaired after a second battery failure nine days later in Japan.

A module controls fans and ducts to expel smoke overboard while keeping an eye out for smoke in the battery area. Low battery voltage caused the power unit to shut down a number of electronics, including ventilation. There was no way to send smoke outside the cabin.

The requirements for the software were all satisfied. However, the requirements were insufficient.

2.6.2.6 Schiaparelli Lander (2016)

A failed Entry, Descent, and Landing Demonstrator Module (EDM) of the ExoMars program, a joint mission of the European Space Agency (ESA) and the Russian Space Agency Roscosmos, was known as Schiaparelli. It was created in Italy with the goal of testing technology for upcoming soft landings on Mars' surface. Additionally, it carried a small but concentrated science payload that would have monitored local climatic conditions and the amount of atmospheric electricity on Mars.

Schiaparelli made an attempt to land on Mars on October 19, 2016, after being launched on March 14, 2016, along with the ExoMars Trace Gas Orbiter (TGO). During the last stages of the landing, Schiaparelli's telemetry signals, which were being tracked in real-time by the Giant Metrewave Radio Telescope in India (and verified by Mars Express), were lost for around a minute from the surface. On October 21, 2016, NASA made a Mars Reconnaissance Orbiter photograph public that appeared to reveal the crash site of the lander. The failure modes of the deployed landing technology were examined using the telemetry data gathered and transmitted by the ExoMars TGO and Mars Express operated by the ESA.

Parachute was deployed at 11 km. Inertial Measurement Uni (IMU) was saturation at 3.7 km; computed negative altitude. Parachute was released; thrusters were turned off. Schiaparelli Lander had collision at 300 km/h (186 mph). It was designed to withstand 10 km/h.

Every component performed according to specification. There were no component failures.

An investigation that concluded in May 2017 identified four root causes for the mishap:

1. Insufficient uncertainty and configuration management in the modeling of the parachute dynamics which led to expect much lower dynamics than observed in flight;
2. Inadequate persistence time of the IMU saturation flag and inadequate handling of IMU saturation by the Guidance Navigation and Control (GNC);
3. Insufficient approach to Failure Detection, Isolation and Recovery, and design robustness;
4. Mishap in management of subcontractors and acceptance of hardware.

According to the board of inquiry's research, the lander started spinning unpredictably quickly right before it deployed its parachute. The guidance, navigation, and control system software made a significant attitude estimation error as a result of Schiaparelli's spin-measuring equipment being briefly saturated by this extremely fast rotation. Due to the computer calculating that it was below ground level, the on-ground system was activated as if Schiaparelli had landed, the parachute and backshell were released early, the thrusters were briefly fired for 3 seconds rather than the normal 30 seconds, and the parachute was released early. The investigation also found that if the persistence time had been chosen at a lower value, the attitude knowledge error brought on by IMU saturation would not have put the mission in danger.

2.6.2.7 Hitomi Satellite (2016)

The Japan Aerospace Exploration Agency (JAXA) commissioned the X-ray astronomy satellite Hitomi, also known as ASTRO-H and New X-ray Telescope (NeXT), to research the most energetic processes in the universe. By examining the hard X-ray band over 10 keV, the space observatory was created to expand the study done by the Advanced Satellite for Cosmology and Astrophysics (ASCA). Originally known as the NeXT, the satellite was launched under the name ASTRO-H. It was given the new name Hitomi after being sent into orbit and having its solar panels extended. The spacecraft was launched on February 17 and lost communication on March 26 as a result of repeated mishaps with the attitude control system, which caused the spacecraft to spin too quickly and fragment.

Hitomi Satellite was unable to detect bright stars for reference. The Safe Hold Mode parameters were incorrect. Every component was performed according to design. It was not a simple component failure.

On April 28, 2016, JAXA gave up trying to recover the satellite and shifted its attention to looking into anomalies. It was found that the spacecraft's inertial reference unit (IRU) reported a rotation of 21.7° per hour around 19:10 UTC on March 25, 2016, even though the vehicle was steady. This started a series of events that eventually led to the loss of the spacecraft. Hitomi rotated in the opposite direction as a result of the attitude control system's attempt to use the reaction wheels of the spacecraft to counteract the spin that never existed. The reaction wheels started to gain excessive momentum as a result of the IRU's continual reporting of inaccurate data, which tripped the spacecraft's computer and caused it to enter "safe hold" mode. The spacecraft's thrusters were then utilized by attitude control to attempt to steady it after the sun sensor failed to lock onto the sun's position and Hitomi began rotating rapidly as a result of an improper software setting. Early on March 26, 2016, due to this erroneous rotation rate, a number of spacecraft components – likely the extended optical bench and solar arrays – broke free.

2.6.2.8 Ford Fusion/Escape (2013): Ford Tells 89,000 Escape, Fusion Owners to Park Cars because of Engine Fire Risk

In a brief period of time (September–December), 13 reports of engine fires were filed. Vehicle parking is requested till further notice for owners. Affected were 99,153 brand-new vehicles.

Under certain conditions, the original cooling system design was unable to resolve a loss of coolant system pressure, which might result in a car fire while the engine was running.

The cooling system software restricted coolant flow as a result of a series of events. That would usually not be an issue and is the intended behavior. However in rare cases, the coolant pressure coupled with other conditions may cause the coolant to boil. When the coolant starts to boil, the engine may experience significant overheating, which accelerates the pressure rise and causes additional boiling. This caused coolant leaks near the hot exhaust that led to an engine fire. Twelve Escapes and one Fusion caught fire.

2.6.2.9 Honda Odyssey: 344,000 Minivans Recalled

Because to a fault with the stability control system, which Honda refers to as Vehicle Stability Assist (VSA), which can result in abrupt, forceful braking without the driver pressing the brake pedal, Honda was recalling 344,000 Odysseys from the 2007 and 2008 model years. Issue was identified in 2013.

This was a software stability control issue. The National Highway Traffic Safety Administration stated that under specific conditions, a software fault may prevent the system from calibrating correctly, causing pressure to build up in the brake system.

The risk of a rear-end collision increases if pressure builds up to a certain degree. The vehicle may suddenly and unexpectedly brake forcefully, and without illuminating the brake lights.

2.6.2.10 Toyota Unintended Acceleration

Toyota has recalled over eight million vehicles in the United States due to two mechanical safety issues: "sticking" accelerator pedals and a design problem that might result in accelerator pedals getting caught in floor mats.

Push-button ignition as of 2004. One hundred two instances of uncontrolled acceleration were reported between 2004 and 2009. Despite stepping on the brake, speeds were higher than 100 mph. Twenty injuries were reported in 30 accidents. Software updates for the pedals and the push-button ignition were required.

2.6.2.11 AI-driven Cart Broken Down on the Tarmac

On the tarmac, a food cart that was AI-driven malfunctioned, circling out of control and getting closer to a gate-parked, vulnerable airplane. The cart was finally brought to a stop by a yellow-vest worker who ran over it with another vehicle and knocked it over. This case has veered off course a bit. The cart was neither mechanically or artificially intelligently controlled in any way. This is regarded as one of the more well-known AI failures.

2.6.2.12 AI Was Unable to Identify Images for Traffic Jam Pilots

The "Traffic Jam Pilot" system, when used by the driver, monitors traffic conditions on the highway and, if it deems it safe to do so, automatically assumes control of the vehicle. It lets the car to follow the one in front of it and maintain its lane position without any driver input.

This is a common instance of an AI failure potentially for Traffic Jam Pilots. The breakthrough in picture recognition, often known as computer vision, launched the triumphal march of deep learning, a group of techniques frequently used to build AI, about 20 years ago. Before going on to more challenging and demanding tasks, it solved previously impossible tasks like differentiating between people and others. The notion that computer vision is a reliable, dependable, and unlikely to fail technology is now universally accepted. Seven thousand five hundred unedited nature photos were acquired by researchers from Berkeley, Chicago, and the University of Washington. These photos baffled even the most advanced computer vision algorithms. Even the most tested and reliable algorithms can occasionally go wrong.

2.6.3 Risk Engineering of Artificial Intelligence: Dealing with Unknown/Unsafe Scenarios

It is important to recognize the risk of dangerously assuming "the issues are known in the field". Unknown and unsafe scenarios are being addressed by Safety of the Intended Functionality (SOTIF), international standard ISO 21448:2022, just published in June 2022. Compliance with this international standard is especially important as AI and ML play key roles in the development of autonomous vehicles. According to the newly published international standard, it is crucial to identify and assess functional insufficiency of AI-based algorithms and the risks associated with their engineering implementations; for example, eye damage from the beam of a Light Detection and Ranging (LIDAR), a remote sensing method that uses light in the form of a pulsed laser to measure ranges (variable distances) to the Earth. Furthermore, the newly published international standard

- addressed "Functional insufficiencies of artificial intelligence-based algorithms" as one of the "safety-relevant topics".
- listed "… when novel technologies (e.g. machine learning) are used or when the Operational Design Domain (ODD) contains a huge space of scenarios, it cannot be claimed that those analyses are sufficient in order to find all relevant insufficiencies and triggering conditions" as one of the sources for initiating unknown and unsafe scenarios.
    - note: ODD stands for Operational Design Domain.

In summary, there is a definite (and urgent) need for my new edition to address the new challenges, development, and opportunities in Risk Engineering and Management. The urgency is partially revealed by the newly published international standard.

Additionally, UL 4600: Standard for Safety for the Evaluation of Autonomous Products is the first safety standard for autonomous vehicle along with other applications and systems. UL 4600 is the first standard designed specifically for Autonomous, Automated, and Connected Vehicles and related products, where AI/ML are playing important roles. Again, it is life-and-death to identify and assess functional insufficiency of AI-based algorithms and the risks associated with their engineering implementations. Dangerously assuming "all AI/ML largely sit in the computer science domain", rather than being part of the engineering one, is extremely risky as I emphasized above. Besides mechanical, civil engineering, so engineering where risk has a larger impact/importance, my new edition will be very beneficial to those whose job titles are related to computer science, computer engineering, and AI/ML engineering among many others such as aerospace engineering, aviation engineering, avionic engineering, safety engineering, reliability engineering, system engineering, mechanical engineering, manufacturing engineering, and industrial engineering. For example,

- In order to develop AI systems that are in line with human needs for mission objectives, the field of research and practice known as "AI Engineering" incorporates the disciplines of systems engineering, software engineering, computer science, and human-centered design.

The section pioneers into the new horizon of Risk Engineering and Management.

TABLE 2.7
Summary of Intentional Failures

Failure Mode Index	Attack	Overview
1	Disturbance attack	Attacker changes the query to obtain the desired response.
2	Attack via poisoning	To obtain the intended outcome, the attacker taints the ML systems' training phase.
3	Model Inversion	Through diligent queries, the attacker is able to obtain the secret features employed in the model.
4	Membership Inference	Attacker can determine whether a specific data record was included in the training dataset for the model or not.
5	Model thievery	An attacker can retrieve the model using skillfully written queries.
6	Reprogramming ML system	Repurpose the ML system to carry out a task for which it was not designed.
7	Adversarial Example in Physical Domain	Attacker uses physical domain to introduce adversarial examples into ML system For instance, 3D printing customized eyewear to trick facial recognition software.
8	A malicious machine learning provider retrieving training data	The customer's training data can be recovered by malicious ML providers by querying the customer's model.
9	ML supply chain being attacked	As the ML models are being downloaded for usage, an attacker hacks them.
10	ML's backdoor	A malicious machine learning provider backdoors an algorithm to turn on when a certain event occurs.
11	Making use of software dependencies	To confuse or manipulate ML systems, the attacker employs common software flaws like buffer overflow.

2.6.4 Failure Modes in Artificial Intelligence (AI)/ Machine Learning (ML)

The need to comprehend how AI/ML systems fail, whether at the hands of an adversary or as a result of a system's inherent design, will only increase in importance as these technologies become more widely used. This section summarizes both of these failure modes in tables.

2.6.4.1 Intentional Failures

Intentional failures are failures where the failure is the consequence of an adversary actively trying to undermine the system in order to achieve her objectives, whether it be to misclassify the outcome, infer personal training data, or steal the underlying algorithm; intentional failures are summarized in Table 2.7.

2.6.4.2 Unintentional Failures

Unintentional failures are failures caused by ML systems that produce formally correct but gravely dangerous results. Unintentional failures are summarized in Table 2.8.

TABLE 2.8
Summary of Unintentional Failures

Failure Mode Index	Failure	Overview
12	Reward Hacking	Due to the discrepancy between the reported reward and the true reward, Reinforcement Learning (RL) systems sometimes behave in unexpected ways.
13	Side Effects	As it works toward its objective, the RL system disturbs the environment.
14	Distributive changes	The system has been tested in a certain setting; however, it cannot adjust to changes in other environments.
15	Examples of Natural Adversaries	Due to hard negative mining, the ML system fails in the absence of attacker perturbations.
16	Common Corruption	Tilting, zooming, or noisy images are examples of common corruptions and disturbances that the system cannot handle.
17	Incomplete Testing	The ML system is not tested in the real-world scenarios where it will actually be used.

BIBLIOGRAPHY

Anderson, T. L. (1995), *Fracture Mechanics: Fundamentals and Application* (2nd ed.), CRE Press, Boca Raton, NY.

Barclay, S. (1970), *The Search for Air Safety: An International Documentary Report on the Investigation of Commercial Aviation Accidents*, William Morrow, New York, NY.

Boehm, B. W. (1991), "Software Risk Management: Principles and Practice," *IEEE Software*, Vol. 8, pp. 32–30.

International Standard Organization (ISO), ISO 21448. (2022), Road vehicles—Safety of the intended functionality.

Lail, P. and Pecht, M. (1995), *Influence of Temperature on Microelectronics and System Reliability: A Physics of Failure Approach*, CRC Press, New York, NY.

Leveson, N. (1991), "Software Safety in Embedded Computer Systems," *Communications of the ACM*, Vol. 34, No. 2, pp. 34–46.

Loss, J. and Kennett, E. W. (1991), "Performance Failures in Buildings and Civil Works," College Park, MD: University of Maryland Architecture and Engineering Performance Information Center. Available from Congressional Information Service (SRI no. 1992U7250-1).

McKaig, T. K. (1963), *Building Failures: Case Studies in Construction and Design*, McGraw-Hill Book Company, New York, NY.

Musa, J. D., lannino, A., and Okumoto, K. (1987), *Software Reliability: Measurement, Prediction, Application*, McGraw-Hill, New York, NY.

Penny, R. K. and Marriatt, D. L. (1979), *Design for Creep*, McGraw-Hill, New York, NY.

Sander, B. (1972), *Fundamentals of Cyclic Stress and Strain*, University of Wisconsin Press, Madison, WI.

Schlager, N. (1994), *When Technology Fails - Significant Technological Disasters, Accidents, and Failures of the Twentieth Century*, Gale Research Inc., Detroit.

Sharirli, M., Butner, J.M., Rand, J. L., McKinney, S. J. and Roush, M. L. (1993), "Probabilistic Risk Assessment for the Worst-case Design-basis Accident at the Los Alamos Neutron Scattering Center," *Proceedings of the International Topical Meeting on Probabilistic Risk Assessment*, Clearwater, FL, pp. 559–564.

Shepherd, Robin and Frost, J. David, eds. (1995), *Failures in Civil Engineering: Structural, Foundation, and Geoenvironmental Case Studies*, American Society of Civil Engineers, New York, NY.

Shooman, M. L. (1983), *Software Engineering: Design, Reliability, and Management*, McGraw-Hill Book Company, New York, NY.

Sundararajan, C. (1995), *Probabilistic Structural Mechanics Handbook: Theory and Industrial Applications*, Chapman & Hall, International Thomson Publishing Inc., New York, NY.

Tarkov, J. (1986), "A Disaster in the Making – American Heritage of Invention and Technology," Spring 1986, pp. 10–17.

UL Enterprise. (2022), "Evaluation of Autonomous Products," UL Standard 4600, Edition 2, Northbrook, Illinois.

Wang, J. X. (1996), "Complexity as a Measure of the Difficulty of System Diagnosis," *International Journal of General Systems*, Vol. 24, No. 3, pp. 257–269.

Wang, J. X. (2017), *Industrial Design Engineering: Inventive Problem Solving*, CRC Press, Boca Raton, FL.

Wang, J. X. (2019a), "Complexity as a Measure of the Difficulty of System Diagnosis in Next Generation Aircraft Health Monitoring System," SAE Technical Paper 2019-01-1357, doi:10.4271/2019-01-1357.

Wang, J. X. (2019b), "A Dynamic Fault Tree Approach for Time-Dependent Logical Modeling of Autonomous Flight Systems," SAE Technical Paper 2019-01-1358, doi:10.4271/2019-01-1358.

3 Risk Assessment
Extending Murphy's Law

3.1 TITANIC: CONNOISSEURS OF ENGINEERING FAILURE

One basis of naval architecture is found in Archimedes' principle, which states that the weight of a statically floating body will equal the weight of the volume of water that it displaces. Floating merely means that the body sinks into the fluid to such a depth that the displaced volume of fluid weighs exactly as much as the whole floating body.

The simplest structural description of a ship is that its hull is a beam designed to support the numerous weights that rest upon it (including its own weight), to resist the local forces produced by concentrated weights and local buoyant forces, and to resist the various dynamic forces that are almost certain to occur. As with any structure, stresses at all points must remain below the limits allowable for the construction material. Likewise, deflections, both local and overall, must be kept within safe limits.

Building a ship that can be neither sunk nor capsized is beyond practicality, but a ship can be designed to survive moderate damage and, if sinking is inevitable, to sink slowly and without capsizing in order to maximize the survival chances of the people aboard.

The most likely cause of sinking would be a breaching of the hull envelope by collision. The consequences of the resulting flooding are minimized by subdividing the hull into compartments by watertight bulkheads. The extent to which such bulkheads are fitted is determined by the International Maritime Organization (IMO) standards that are based on the size and type of ship. At a minimum, ships that must have a high probability of surviving a collision (e.g., passenger ships) are built to the "one-compartment" standard, meaning that at least one compartment bounded by watertight bulkheads must be floodable without sinking the ship. A two-compartment standard is common for larger passenger-carrying ships – a measure that presumably protects the ship against a collision at the boundary between two compartments. The *Titanic*, the victim of the most famous sinking in the North Atlantic, was built to the two-compartment standard.

Titanic's keel (number 401) was laid down on March 31, 1909 and was launched on May 31, 1911. This giant ship had nine decks. Below the topmost boat deck were decks A, B, C, D, E, F, and G. Below the G deck were the boiler rooms and holds. As shown in Figure 3.1, the hull was divided into 16 watertight compartments by means of 15 watertight bulkheads extending up to deck F. Heavy watertight doors provided communication between bulkheads during normal operations; these were electrically operated and also had a floatation mechanism for automatic closing.

DOI: 10.1201/9781003371014-3

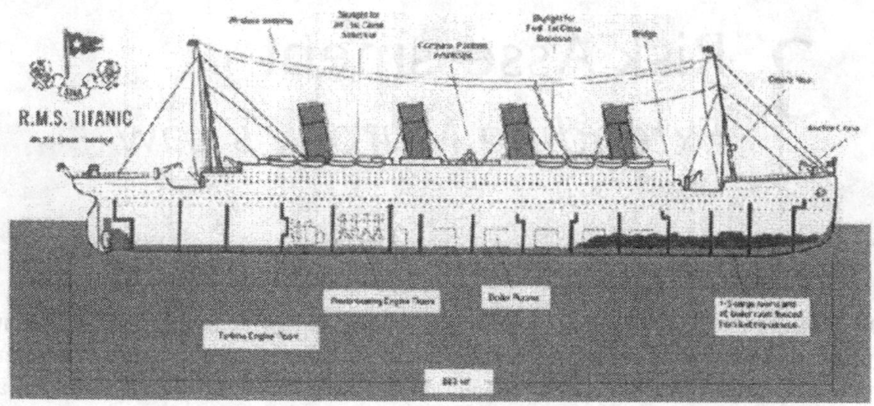

FIGURE 3.1 Titanic: "unsinkable" with its 16 watertight compartments.

Titanic had a double-bottomed hull that was divided into 16 presumably watertight compartments. She was designed to remain afloat with any two compartments flooded, possibly four, enabling her to withstand a collision at the joint of two compartments. This was the worst disaster anyone at that time could have imagined. It was widely regarded that the "Olympic Class" liners were largely unsinkable and were themselves life boats.

Shortly before midnight on April 14, the ship collided with an iceberg (see Figure 3.2); six of its watertight compartments were ruptured, causing the ship to sink at 2:20 am April 15, 2 hours and 40 minutes after the collision.

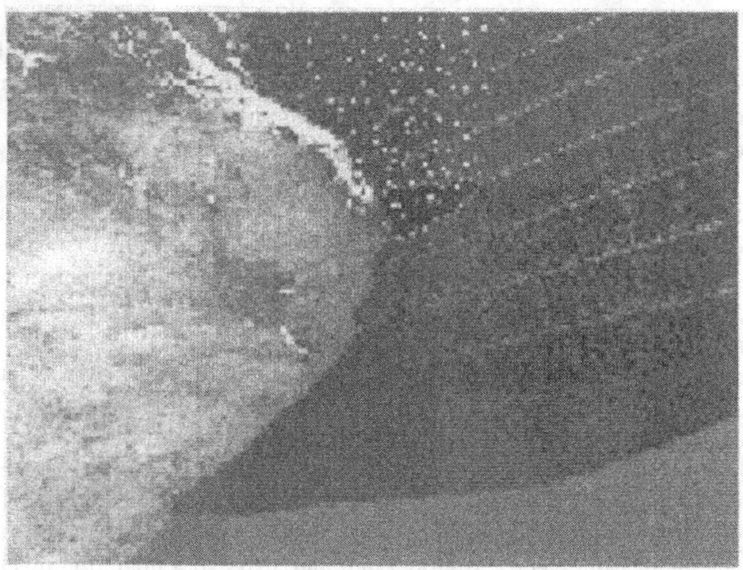

FIGURE 3.2 Titanic: Struck by an iceberg.

New research indicates the ship foundered after suffering six narrow gashes, not one large one, as thought. Computer calculations help reveal the detailed stages:

- When the Titanic struck the iceberg, six of its 16 watertight compartments were damaged. The ship started taking on water in the bow through openings about 20 feet below the water line.
- As the liner nosed down, water flooded compartments one after another and the ship's stern began rising out of the water.
- As the stern rose ever higher, the stress amidships was more than the vessel could bear. It broke apart just forward of the third funnel. The bow began to sink.
- With the bow gone, the stem temporarily settled back to nearly level. The stem then rapidly flooded and rose out of the water, pivoting on the surface. It held that vertical position about a minute and then slowly slipped beneath the surface.
- The bow section sank gently and plowed into the mud. The stem's impact was more traumatic: when it hit bottom, it buried itself some 50 feet, crumpling the steel hull.

Why did the unsinkable Titanic sink within only 2 hours and 40 minutes and cause 1,523 causalities? Eighty years after its Atlantic sinking, the Titanic's true failure mode was still open to question.

Scaling Effect with Ship Design: Since no large gash was found on the recently discovered wreck, researchers indicated that the mere size of the great ship may have as much to do with its failure as the sharpness of the iceberg. Because small- and medium-sized metal structures had proved to be safe in service, engineers and architects have scaled them up, assuming that if the stress remained the same, the large structure would be as safe as the smaller one. The scaling effect may have influenced the ship's damage buoyancy after a collision.

Materials Under Strength: In 1993, a team of architects and engineers released a report in which they argued that the tragedy was caused not so much by the collision with the iceberg as by the structural weakness of the ship's steel plates. Low-grade steel such as that used on the Titanic is subject to brittle fracture rather than bending in cold temperatures. If a better grade of steel had been used, the ship might have withstood the collision or, at the very least, sunk more slowly.

The February 1995 edition of Popular Mechanics had an article about a metal test on a chunk of what they believed to be the hull of the Titanic about the size of a Frisbee with three rivet holes. They used the Charpy test on it. A slice of this hunk was placed in a bath of –1°C alcohol along with a slice of good grade standard ship steel. The slice was tested first with a swinging 67 pound weight that bent it into a V shape. The Titanic slice shattered upon impact and fractured pieces flew in the air. The Titanic engineers did have strong steel, but for stress, not for brittleness. The high sulfur content of this sample was the reason for this metal being low grade and brittle. It was even compared to a rivet slug that was taken from the ship yard in Belfast and was shown to have not changed physically (cell structure) in 80 years.

Recent research indicates that the luxury liner Titanic, the biggest ship of its day, may have gone to the depths because of some of its smallest parts: its rivets. Two wrought-iron rivets salvaged from the ship were found to contain high concentrations of slag, which would have made them dangerously brittle. The 46,000-ton RMS Titanic was held together by about three million rivets that secured its steel beams and side plates. Each rivet was an inch in diameter and 2–3 inches long. Those rivets were unable to hold the plates together after the ship struck the iceberg.

Human Error: The Titanic had received a number of warnings of ice in the region, and although Captain E. J. Smith is known to have seen at least four of these, he did not alter his speed of about 21 knots. On the night in question, the pair of look-outs were working without binoculars, which were supposed to be standard equipment on the White Star Line. They did not see the iceberg until it was only a quarter mile away. The bow was swung swiftly, but it was too late.

The order to ready the lifeboats was given 40 minutes after the collision. Since there had been no boat drill, the crew and passengers did not know which lifeboats they should board. Due to organizational errors during the boat loading process, many boats went away only partly filled, with room for about 467 more people.

The mishaps in wireless communication also contributed to further unnecessary loss of life on the Titanic. The wireless operator on the Californian, a ship not more than 20 miles away, did not follow around-the-clock shipboard watches. He had stopped working only 15 minutes before the Titanic's wireless operator placed a distress call.

So many things could go wrong. The puzzle of the Titanic sinking illustrates the need to extend Murphy's Law: "if anything can go wrong, we should know how likely it will be to go wrong". With many failure scenarios, we must direct our resources wisely to reduce our exposure to risk.

3.2 RISK ASSESSMENT: "HOW LIKELY IT IS THAT A THING WILL GO WRONG"

One of the principal aims of engineering design is the assurance of system performance within the constraint of economy. The achievement of this objective, however, is generally not a simple problem, particularly for large engineering systems. Engineering systems occasionally fail to perform their intended function, such as in the case of the sinking of the "unsinkable Titanic". From this perspective, risk is generally inherent in all engineered systems.

Risk assessment is concerned with quantities, e.g., how many units are in use? How many parts? How many fail? What proportion is this to the total? Therefore, statistical analysis can be important in analyzing these quantities. Statistical analyses are also an essential tool in calculating probabilities of failure. Mathematically, "probability" can be expressed in three ways, all of which in practice mean the same thing.

1. As a percentage, e.g., 99%;
2. As odds, e.g., 99:1;
3. As a decimal fraction, e.g., 0.99.

What most users want to know is an equipment's probability of failure over a given period of time and the consequence of failure. If we start with the equipment working and determine the likelihood of failure by some later time, we find by practical experience a behavior that can be represented by some form of mathematical equation. If the equipment has constant likelihood of failure per unit of time, this equation is exponential in form. This indicates that there are few failures at the beginning, but the cumulative probability of failure increases as time goes on.

The probability of failure occurring by a given time can be expressed as follows:

$$P_f = 1 - e^{-\lambda t}$$

where

λ is the constant failure rate;
P_f *is* the probability of failure by time t;
e is the base of the natural system of logarithms and equals 2.718.

For small λt,

$$P_f = 1 - e^{-\lambda t} \cong \lambda t$$

The concept of probability is fundamental to an understanding of what is meant by risk, so it is vital to be quite clear of what "probability" means.

If we have a deck of 52 well-shuffled cards, the first card drawn from the deck has a probability of 1 in 52 (0.02) of being the ace of spades. After we select the first card, we know what the card is and the probability of it being the ace of spades is then either 1 or 0. If we plan to select 26 cards at random from the complete deck, the probability that the ace of spades will be included in those selected is 26 of 52, or 0.5. If we were to repeat these trials many thousands of times, the fraction of trials in which the ace of spades was selected would approximate the value that we indicated was the probability of occurrence for a single trial.

In risk engineering, the likely period for which an item will function before failure is most often represented by the mean time between failures (MTBF). But there is no certainty that a given item will not fail before the end of this period; it might fail within a much shorter period, or perhaps run for longer without failure. But as determinations of the time to failure are made for more and more items, the measured MTBF will more accurately represent the basic property of these items. For items with constant failure rate, the MTBF can be calculated as follows:

$$\text{MTBF} = \frac{1}{\lambda}$$

or

$$\lambda = \frac{1}{\text{MTBF}}$$

then

$$P_f = 1 - e^{\frac{-t}{\text{MTBF}}}$$

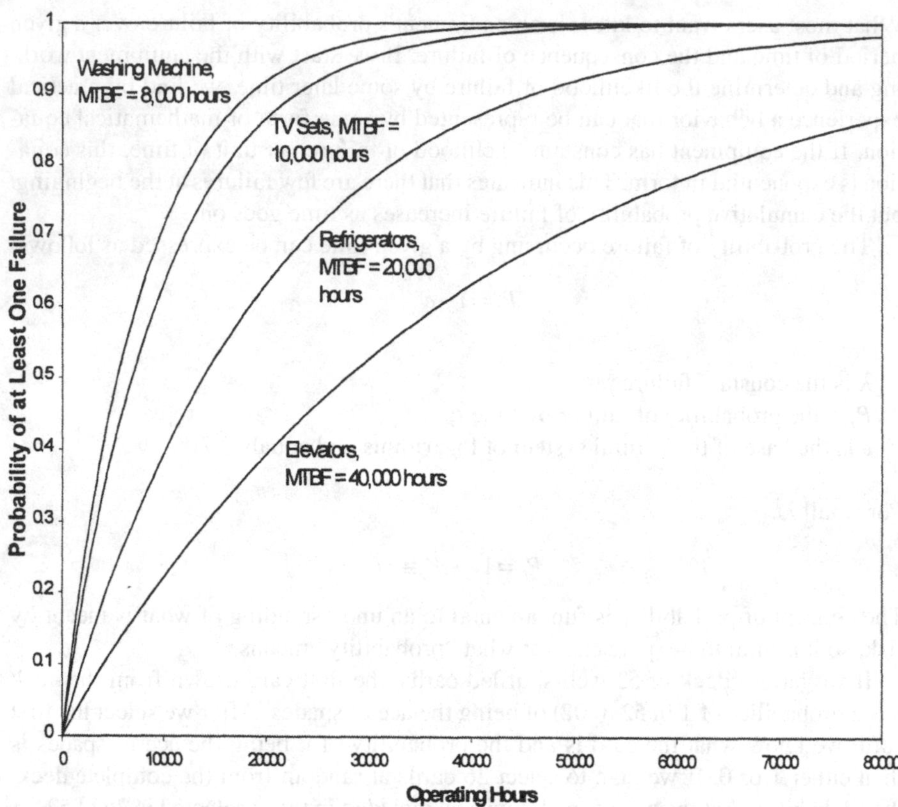

FIGURE 3.3 Failure probability vs. operating hours.

It is important to remember that one cannot calculate the exact period for which a specific item of equipment will work without failure. All that can be done is to calculate the probability of an equipment working for a particular period of time (see Figure 3.3).

Unless you have an item that fails frequently, difficulty arises in estimating failure rate (or MTBF) statistically because of the very large number of components which must be tested to get a meaningful number of failures. For instance, if 330,000 specimens of a component with a failure rate of 0.001% per 1,000 hours were tested for 1,000 hours (nearly 42 days), there would be only three failures on average. It is unlikely that the production run would be large enough to justify testing on this scale, and the time, equipment, and expense involved are considerable.

For high reliability items such as integrated circuits, a new approach has therefore been developed, termed the physics of failure approach, which sets out to prevent the manufacturing of faulty components, in contrast to detecting and rejecting them. The integrated circuit itself has become extremely reliable; faults can occur in encapsulating it, mounting it, and connecting it to lead-out wires. Through careful control of processes, faults in these areas can be avoided.

For real equipment, the failure rate, λ, may not be constant. When a new purchase is made, whether it is a radio, a washing machine, an aerospace system, or a tank, early failures may occur. These frequently are caused by manufacturing faults or misuse. The early failure rate may, therefore, be relatively high, but the rate falls as the weak parts are replaced. There follows then a period during which the failure rate is relatively low and fairly constant, and finally the failure rate rises again as parts start to wear out. For a non-constant failure rate, $\lambda(t)$, the probability of failure by time t can be calculated as:

$$P_f(t) = 1 - e^{-\int_0^t \lambda(t)dt}$$

Most planning and design of engineering systems must be accomplished without the benefit of complete information; consequently, the assurance of performance can seldom be perfect. Nevertheless, many decisions that are required during the process of planning and design are invariably made under conditions of uncertainty. Therefore, there is invariably some chance of non-performance or failure and of its associated adverse consequences; hence, risk is often unavoidable. In the case of a structure, its risk is clearly a function of the maximum load (or combination of loads) that might be imposed over the useful life of the structure. Structural risk will also depend on the strength or load-carrying capacity of the structure or its components. Since the accurate lifetime maximum load and the exact capacity of a structure are difficult to predict, and any prediction is subject to uncertainties, the absolute assurance of zero-risk is impossible. Realistically, low-risk may be assured only in terms of the probability that the available strength (or structural capacity) will be adequate to withstand the lifetime maximum load. In the next section, the risk assessment method will be extended to systems involving multiple failure modes.

3.3 RISK ASSESSMENT FOR MULTIPLE FAILURE MODES

Engineering problems often involve multiple failure modes; that is, there may be several potential modes of failure, in which the occurrence of any one of the potential failure modes will constitute failure or malfunction of the system or component. For example, a structural element may fail in flexure or shear or buckling or their combinations. Similarly, for a multi-component structural system, failures of different sets of components may constitute different failure modes.

We will introduce selected mathematical operators to assist in the following discussion. The union and intersection operators provide two basic logical relationships among quantities and may be used with a set of multiple failure modes. Considering a building foundation, failure may be caused by inadequate bearing capacity (E_1) or excessive settlement (E_2). The failure of the building foundation can be expressed as the union of events E_1 and E_2, the occurrence of E_1 or E_2, or both:

$$\text{Building foundation failure} = E_1 \cup E_2$$

In the case of an environmental system, suppose two methods are provided to control a major pollutant; inadequate control of the pollutant will occur only upon failure of the first control method (E_1) and the second control method (E_2). This can be expressed as the intersection of the events E_1 and E_2, the occurrence of both E_1 and E_2:

$$\text{Inadequate control of the pollutant} = E_1 \cap E_2$$

It is obvious that the more parts in any machine or equipment, the more risk that there will be at least one failure. The risk assessment of a multi-component system is essentially a problem involving multiple modes of failure; that is, the failures of different components or different sets of components constitute distinct and different failure modes of the system. The consideration of multiple modes of failure, therefore, is fundamental to the problem of system risk assessment. The identification of the individual failure modes and the evaluation of the respective probabilities may be problems in themselves.

Consider a system with k potential failure modes. Let E_i be the event of failure by the ith failure mode.

The event of non-failure by the ith mode is noted by

$$\bar{E}_i$$

The probability of failure of the ith mode is

$$P_i = P(E_i)$$
$$1 - P(\bar{E}_i)$$

or

$$P(\bar{E}_i) = 1 - P_i$$

The event of failure of the system, E, is the event that one or more of the failure modes occur

$$E = E_1 \cup E_2 \cup \ldots \cup E_k$$

The probability of system failure, P(E), is calculated by

$$\begin{aligned} P(E) &= P(E_1 \cup E_2 \cup \ldots \cup E_k \\ &1 - P(\bar{E}_1 \cap \bar{E}_2 \ldots \cap \bar{E}_k) \end{aligned} \tag{3.1}$$

Independence will be true if the probability of failure of any mode is not influenced by that of any other mode. In the case of independence,

$$P(\bar{E}_1 \cap \bar{E}_2 \ldots \cap \bar{E}_k) = P(\bar{E}_1) P(\bar{E}_2) \ldots P(\bar{E}_k) \tag{3.2}$$

so,

$$P(E)=1-P(\bar{E}_1\cap\bar{E}_2\ldots\cap\bar{E}_k)$$
$$1-P(\bar{E}_1)P(\bar{E}_2)\ldots P(\bar{E}_k) \quad (3.3)$$
$$1-(1-P_1)(1-P_2)\ldots(1-P_k)$$

Example 3.1

A mechanical system consists of a gear, a shaft, and two bearings.

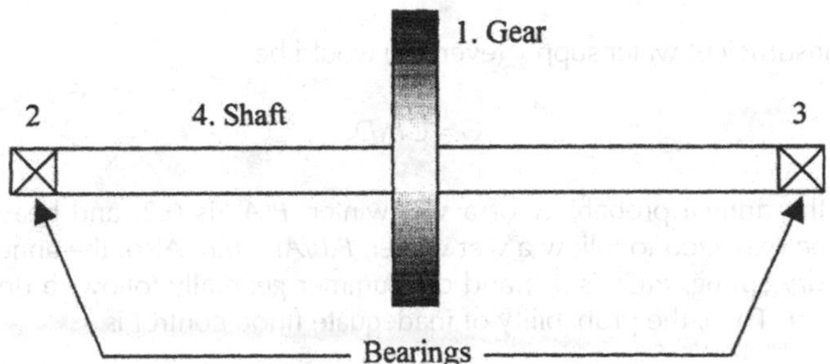

The probabilities of component failures are as follows:

1. Gear $P_1 = 5E-4$;
2. Bearings $P_2 = P_3 = 2.5E-4$;
3. Shaft $P_4 = 5E-5$.

The probability of failure of the mechanical system can be calculated as follows:

$$\begin{aligned}P(E)&=1-(1-P_1)(1-P_2)(1-P_3)(1-P_4)\\&=1-(1-0.0005)(1-0.00025)^2(1-0.00005)\\&=1.05E-3\end{aligned}$$

Generally, the different failure modes may be correlated. The system failure probability can then be calculated by

$$\begin{aligned}P(E)&=P(E_1\cup E_2\cup\ldots\cup E_k)\\&=P(E_1)+P(E_2)+\ldots+P(E_k)\\&-P(E_1\cap E_2)-P(E_1\cap E_3)-\ldots-P(E_{k-1}\cap E_k)\\&+P(E_1\cap E_2\cap E_3)+P(E_1\cap E_2\cap E_4)+\ldots P(E_{k-2}\cap E_{k-1}\cap E_k)\\&+(-1)^{K+1}P(E_1\cap E_2\ldots\cap E_k)\\&\le P(E_1)+P(E_2)+\ldots+P(E_k)\end{aligned} \quad (3.4)$$

Example 3.2

For example, a reservoir may be designed for both flood control and water supply. Flooding may occur only in the spring caused by the melting of heavy accumulations of snow in the previous winter (event A) coupled with abundant rainfall in the spring (event B). However, water supply may be inadequate only in the summer and fall; water shortage may occur if the reservoir water is low in early summer (event C) because of a dry winter or spring coupled with a low rainfall over the summer (event D).

Inadequate flood control (event F) is then

$$F = A \cap B$$

whereas insufficient water supply (event G) would be

$$G = C \cap D$$

Suppose the annual probability of a wet winter, $P(A)$, is 0.2; and heavy spring rain may be expected to follow a wet winter, $P(B/A) = 0.8$. Also, the annual probability of dry spring, $P(C)$, is 0.1; and dry summer generally follows a dry spring, $P(D/C) = 0.9$. Then, the probability of inadequate flood control is

$$\begin{aligned} P(F) &= P(A \cap B) \\ &= P(A)P(B/A) \\ &= 0.2 \times 0.8 \\ &= 0.16 \end{aligned}$$

Similarly, the probability of insufficient water supply is

$$\begin{aligned} P(G) &= P(C \cap D) \\ &= P(C)P(D/C) \\ &= 0.1 \times 0.9 \\ &= 0.09 \end{aligned}$$

According to Equation (3.4), the probability of unsatisfactory reservoir performance is

$$\begin{aligned} P(E) &= P(F \cup G) \\ &\leq 0.16 + 0.09 \\ &\leq 0.25 \end{aligned}$$

One could study the influence on this probability of reservoir size. By increasing the size of the reservoir, one could reduce both components of reservoir failure. Of course, it would be necessary to do a cost benefit study for such a size increase.

In a complex multi-component engineering system, the possibilities of failure or the different ways in which failure of the system can occur may be so involved that a systematic scheme for identifying all the potential failure modes and their respective consequences is necessary. The fault tree analysis (FTA) (see Section 3.4) and event tree analysis (see Section 3.5) serve this purpose.

3.4 FAULT TREE ANALYSIS: DEDUCTIVE RISK ASSESSMENT

For a complex system, the adverse conditions or faults that could lead to the occurrence of system failure may be quite involved. FTA is a risk assessment technique which starts from consideration of specific system failure events, referred to as "top events". The analysis proceeds by determining how these can be caused by individual or combined lower level failures or events. This "deductive" approach starts with the undesired final outcome and analyzes backward to determine how this result could come about. This approach is sometimes likened to that of Sherlock Holmes who started with a dead body and determined what happened to produce this death.

An FTA may include a quantitative evaluation of the probabilities of the various faults or failure events leading eventually to the calculation of the probability of the top event. Used in a quantitative form, FTA is a valuable diagnostic tool. Aside from identifying all potential paths that could lead to failure, it can also serve to single out the critical events that contribute significantly to the likelihood of failure of a system and reveal potential weak links in the system. Finally, the fault tree can be used to calculate the demand failure probability, system unreliability, or unavailability of the system in question. This task of quantitative evaluation is often of primary importance in determining whether risk associated with a final design is acceptably low.

A fault tree diagram is a graphical decomposition of a top event into the union and/or intersection of sub-events. The alternative faults that could lead to the top event are logically related to the top event by "OR" gates and "AND" gates. An "OR" gate indicates that the top event is the union of sub-events, i.e., the occurrence of any one of these sub-events will cause the top event. An "AND" gate signifies that the top event is the intersection of sub-events, i.e., the top event only occurs if all sub-events occur.

Example 3.3

"OR" Gate: An automotive brake system consists of master cylinder, brake lining, wheel cylinder, and brake fluid.

The central component of the brake system is the master cylinder (see Figure 3.4a). When the brake pedal is depressed, the force on the piston produces a pressure in the chamber behind it. This is transmitted by the brake fluid to the wheel cylinders.

The "top event" for this system is simply labeled as "Brake System Failure".

From the statement of system operation, one may conclude that the system will fail in case of the occurrence of any one of the following events:

- Master cylinder fails to produce required pressure (*El*);
- Insufficient brake fluid to transmit the pressure to wheel cylinder (*E2*);

FIGURE 3.4a A master cylinder and brake equipment.

- Wheel cylinders fail to provide adequate braking ($E3$);
- Brake lining fails to provide adequate braking ($E4$).

The probability of the top event can be calculated by

$$P(T) = 1 - [1 - P(E1)][1 - P(E2)][1 - P(E3)][1 - P(E4)]$$

If the probabilities of the individual events are small, then

$$\approx P(E1) + P(E2) + P(E3) + P(E4)$$

Example 3.4

"AND" Gate: The brake system of an automobile consists of independent front and rear brakes, i.e., two independent sets of the brake equipment, each similar to that illustrated in Example 3.1.

An accident caused by the brake failure of this automobile requires a combination of the following events:

- the failure of the front brake (FB);
- the failure of the rear brake (RB);
- the inability of the driver to take proper evasive action (H).

The fault tree for the automotive brake system, including the driver's failure in emergency response, is shown in Figures 3.4b and 3.5. The probability of the top event can be calculated as follows:

$$P(T) = P(\text{FB}) \times P(\text{RB}) \times P(\text{H})$$

Each of the sub-events can be decomposed further until the sequence of events leads to basic causes of interest, called "basic events". The "basic events" are usually those whose occurrence probabilities can be readily assessed. The system discussed in Example 3.2 is a redundant system, since either FB or RB can stop the car. The sub-events FB and RB could be decomposed further as shown in Example 3.1.

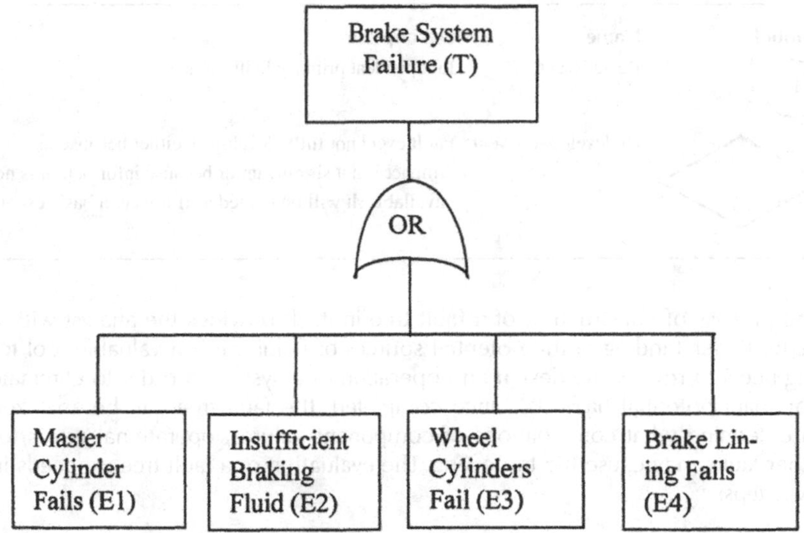

FIGURE 3.4b Fault tree for an automotive brake system.

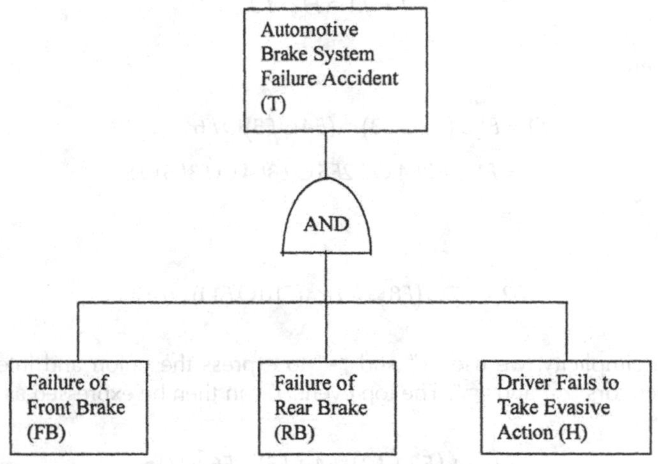

FIGURE 3.5 Fault tree for an automotive brake system failure accident.

Example 3.5

Completing a fault tree for the automotive brake system by combining fault trees of Figures 3.4b and 3.5. A wheel cylinder failure might occur in the left or right wheel.

The fault tree representation of the system is shown in Figure 3.5. The following symbols are used in the fault tree.

Symbol	Name	Description
(E)	Basic Event	Independent primary fault event.
◇E	Undeveloped Event	Fault event not fully developed, either because its impact is not significant or because information is not available. It will be treated as if it were a basic event.

The process of construction of a fault tree in itself provides the analyst with a better understanding of the potential sources of failure. It is a valuable tool for engineers to review the design and operation of a system in order to eliminate dominant potential hazards. Once completed, the fault tree can be analyzed to determine what combinations of component failures, operational errors, or other faults may cause the top event. The evaluation of a fault tree proceeds in two steps:

1. *Qualitative Analysis*:– A logical expression is constructed for the top event in terms of combinations of basic events; a bottom-up approach is shown in Figures 3.6a and 3.6b, from which we find that

$$T = T1 \cap H \cap T2 \tag{3.1}$$

where

$$\begin{aligned} T1 &= E1 \cup (E2 \cup E3) \cap (E4 \cup E5) \cup E6 \\ &= E1 \cup E2E4 \cup E2E5 \cup E3E4 \cup E3E5 \cup E6 \end{aligned} \tag{3.2}$$

and

$$T2 = E7 \cup (E8 \cup E9) \cap (E10 \cup E11) \cup E12$$

For simplicity, we use "+" and "•" to express the union and intersection operators "∩" and "∩". The top event, *T*, can then be expressed as follows:

$$T\left[E1 + (E2 + E3)(E4 + E5) + E6\right](H) \bullet$$

The minimal cut set is an important concept for FTA. A minimal cut set is defined as the smallest combination of basic events which, if they all occur, will cause the top event to occur. For example, $H \cap El \cap E7$ is a combination of basic events sufficient to cause the top event. If even one of the failures in the minimal cut set, *H*, *El*, or *E7*, does not happen, the top event will not take place due to that cut set. So, $H \cap El \cap E7$ is a minimal cut set. Knowing the minimal cut sets for a particular fault tree can provide valuable insight concerning potential weak points of complex systems.

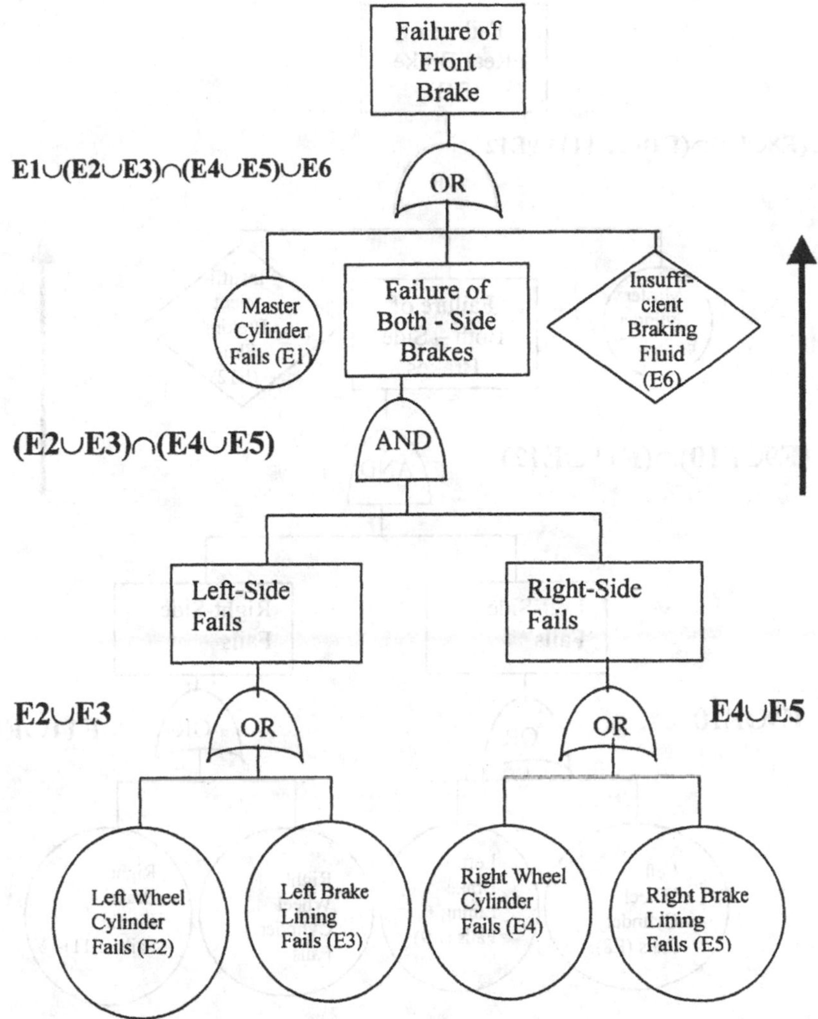

FIGURE 3.6a Fault tree and logic expression representing failure of front brake.

The minimal cut sets for this fault tree are given in Table 3.1. If the probability of occurrence for each of the basic events is small compared with unity, then the cut sets with more terms will be lower in probability of occurrence. Thus, a qualitative review of our fault tree leads us to focus first on the four cut sets that have only three basic events required for them to occur. This qualitative analysis provides us first a list of the minimal cut sets which we can review to make sure that we really do understand the combinations of circumstances which will result in system failure. In addition, we roughly order these cut sets by the number of terms and quickly focus on the cut sets that are likely to dominate.

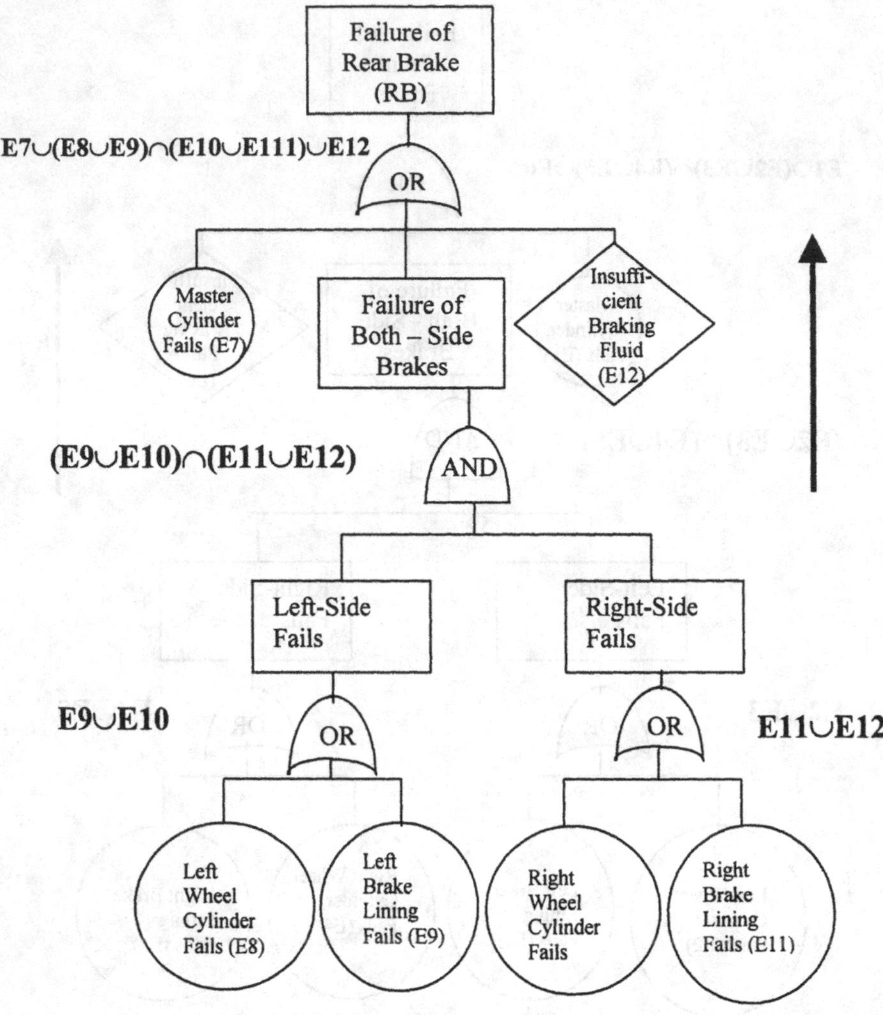

FIGURE 3.6b Fault tree and logic expression representing failure of rear brake.

2. *Quantitative Analysis*: – The logical expression can be used to give the probability of the top event in terms of the probabilities of the primary events (Figure 3.6c). Suppose the basic events for the automotive brake system have the following annual probabilities of occurrence:

Basic Events	Description	Annual Probability of Occurrence
*E*1,*E*7	Master Cylinder Failure	0.01
*E*2, *E*4, *E*8, *E*10	Wheel Cylinder Failure	0.02
*E*3, *E*9, *E*5, *E*ll	Brake Lining Failure	0.05
*E*6, *E*12	Insufficient Braking Fluid	0.02
H	Driver fails to take evasive action	0.3

TABLE 3.1
Minimal Cut Sets for the System in Figures 3.5 and 3.6c

	4 3-term Cut Sets	16 4-term Cut Sets	16 5-term Cut Sets
1	E1•H•E7	E1•H•E8•E10	E2•E4•H•E8•E10
2	E1•H•E12	E1•H•E8•E11	E2•E4•H•E8•E11
3	E6•H•E7	E1•H•E9•E10	E2•E4•H•E9•E10
4	E6•H•E12	E1•H•E9•E11	E2•E4•H•E9•E11
5		E6•H•E8•E10	E2•E5•H•E8•E10
6		E6•H•E8•E11	E2•E5•H•E8•E11
7		E6•H•E9•E10	E2•E5•H•E9•E10
8		E6•H•E9•E11	E2•E5•H•E9•E11
9		E2•E4•H•E7	E3•E4•H•E8•E10
10		E2•E5•H•E7	E3•E4•H•E8•E11
11		E3•E4•H•E7	E3•E4•H•E9•E10
12		E3•E5•H•E7	E3•E4•H•E8•E11
13		E2•E4•H•E12	E3•E5•H•E8•E10
14		E2•E5•H•E12	E3•E5•H•E8•E11
15		E3•E4•H•E12	E3•E5•H•E9•E10
16		E3•E5•H•E12	E3•E5•H•E9•E11

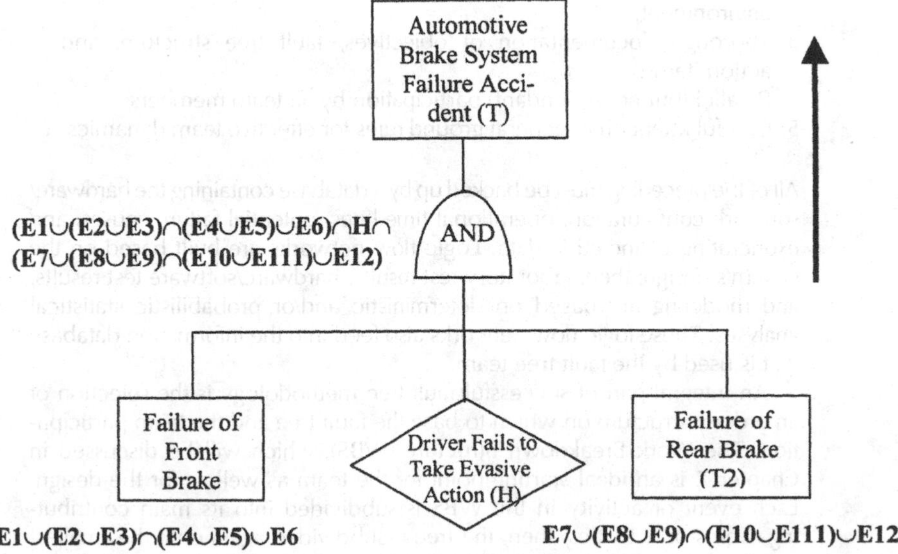

FIGURE 3.6c Complete logic expression of top event.

From Equation (3.2), assuming independence of events:

$$P(T1) = P\big(E1 \cup (E2 \cup E3) \cap (E4 \cup E5) \cup E6\big)$$

Similarly,

$$P(T2) = P\big(E7 \cup (E8 \cup E9) \cap (E10 \cup E11) \cup E12\big)$$

Because $T1$ and $T2$ are independent, we can evaluate the probability of each alone and then use these probabilities in Equation (3.1) obtaining:

$$P(T) = P(H \cap T1 \cap T2)$$

To engineers, the process of FTA is of major value in identifying and investigating potential failure modes and related causes. FTA can be applied in the early design phase of a system and then progressively refined and updated as the design evolves. It is very helpful in obtaining knowledge about the system design with respect to potential failures. FTA can be used as a tool to develop robust engineering designs.

FTA is a systematic approach to fault evaluation achieved by postulating potential high-level faults and identifying the primary and secondary causes, down to the lowest piece-part, that could induce the high-level fault. In situations of high urgency and cost or schedule sensitivity, it is often desirable to apply a team approach to development and use of the fault tree methodology, such as the one shown in Figure 3.7.

The keys to a successful team approach to FTA are as follows:

1. Selection of the right people, such as responsible reliability engineers, design engineers, and system engineers to participate in the analysis;
2. Interactive meetings of these people in a creative but focused environment;
3. Thorough documentation of objectives, fault tree structure, and action items;
4. Parallel (but not redundant) participation by all team members;
5. Careful attention to general ground rules for effective team dynamics.

All of the preceding must be backed up by a database containing the hardware/software configuration, operational time lines, potential failure causes, and exonerating or indicting data. Logic flow networks are built based on the system's design; then, laboratory test results, hardware/software test results, and modeling are based on deterministic and/or probabilistic statistical analyses. These logic flow networks also feed into the information database that is used by the fault tree team.

An integral part of successful fault tree methodology is the selection of an orderly structure on which to base the fault tree and the team participation. The Work Breakdown Structure (WBS), which will be discussed in Chapter 7, is an ideal starting point for the team as well as for the design. Each event or activity in the WBS is subdivided into its main contributing events or activities; then, the tree is subdivided again until the smallest activity that cannot be further subdivided is reached. These lowest events or activities are the "leaves" of the fault tree.

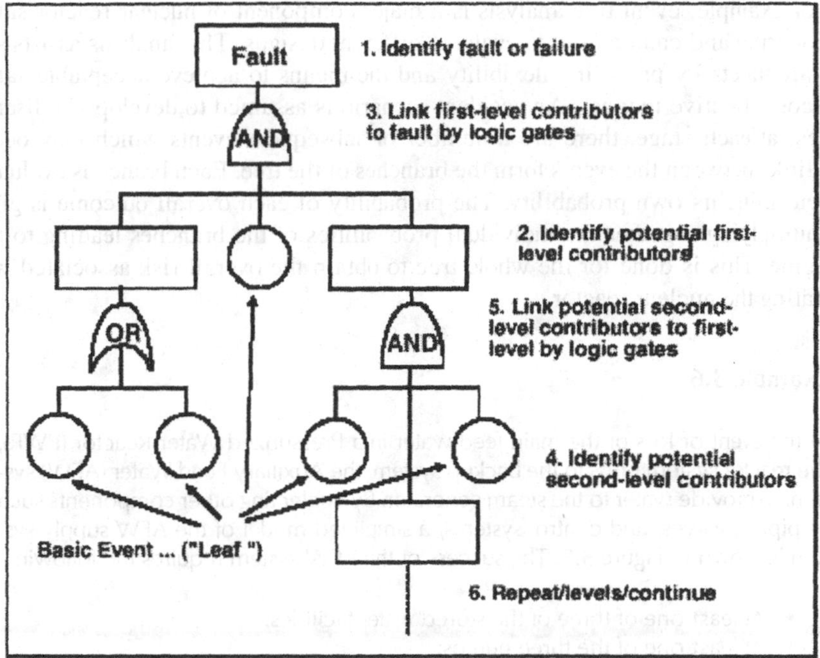

FIGURE 3.7 A fault tree team methodology (NASA preferred reliability practice no. PD-AP-1312).

3.5 EVENT TREE ANALYSIS: INDUCTIVE RISK ASSESSMENT

In many accident scenarios, the initial local failures, called initiating events, have a wide spectrum of results, ranging from inconsequential to catastrophic. The consequences are determined by how the accident progression is affected by subsequent failure or operation of other components or subsystems, particularly safety or protection devices, and by human errors made in response to the initiating event. In such situations, an inductive method may be very useful. We begin by asking "what if" the initiating event occurs and then follow each of the possible sequences of events that result from assuming failure or success of the components and humans affected as the accident propagates. After such sequences are defined, we may attach probabilities to them if such a quantitative estimate is needed.

The event tree is a useful technique for such an inductive analysis, whether used in a qualitative or quantitative manner. Event trees are very useful in analyzing the effects of the functioning or failure of safety systems in response to an accident, particularly when events follow with a particular time progression. In general, whether a particular consequence of an initiating event is serious or not may depend on whether a sequence of other adverse events occurs following the initiating event. In order to ascertain that all potentially dangerous or adverse consequences are considered following an initiating event, the different possible sequences of subsequent events are methodically identified by an event tree analysis.

For example, event tree analysis is a major component of nuclear reactor safety engineering and can be used to evaluate different designs. This analysis assists system architects by providing flexibility and the means to achieve acceptable safety in a cost-effective manner. An accident scenario is assumed to develop via discrete stages; at each stage, there are a number of subsequent events which may occur. The links between the events form the branches of the tree. Each branch is evaluated to determine its own probability. The probability of each overall outcome is given by multiplying together the individual probabilities of the branches leading to that outcome. This is done for the whole tree to obtain the overall risk associated with operating the nuclear reactor.

Example 3.6

In the event of loss of the main feed water in a Pressurized Water Reactor (PWR), the reactor system turns to the backup system, the Auxiliary Feed Water (AFW) system, to provide water to the steam generator. By neglecting other components such as piping, valves, and control systems, a simplified model of the AFW supply system is shown in Figure 3.8. The success of the AFW system requires the following:

- At least one of three of the stored water facilities;
- At least one of the three pumps.

One of the pumps is steam-turbine-driven, whereas the other two are electric pumps that depend on the functioning of electric power supply from diesel generators.

We are concerned here with the following events:

- The initiating event, *IE*, is the loss of main feed water supply; its probability is denoted as *P*(*IE*);

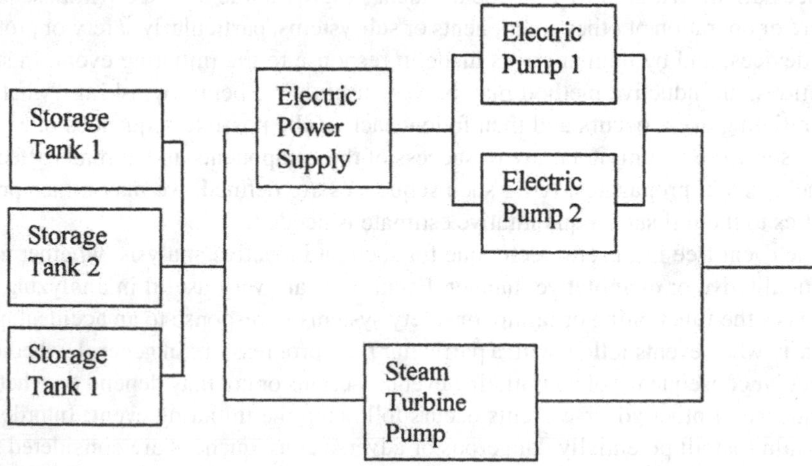

FIGURE 3.8 Functional reliability diagram for auxiliary feed water system in a PWR.

TABLE 3.2
Event Tree Branch Probabilities

Event	Description	Probability of Occurrence
IE	Loss of Main Feed Water	0.2 per year
El	No Water Supply from Three Storage Tanks	*lE*-5
E2	Failure of both Electric Pumps, including failure caused by malfunction of the electric power supply	2*E*-3
E3	Failure of Turbine Pump	*lE*-2

- The following event, *El*, is the supply of water from the storage tanks to the pumping systems; its probability of failure is denoted as *P*(*E*1);
- The second event, *E*2, is the output from at least one of two electric pumps; its probability of failure is denoted as *P*(*E*2); event *E*2 can be further broken down into:
  1. Failure of both electric pumps;
  2. No electric power supply;
- The third event is the output from the turbine pump; its probability of failure is denoted as *P*(*E*3).

Assume the following probability data for the initiating event and the following system responses (Table 3.2).

An event tree for the loss of main feed water is developed (see Figure 3.9). The probability of failure can be calculated accordingly:

$$P(\text{Failure}) = P(IE) * P(E2) * P(E3) + P(IE) * P(E1)$$
$$= 4E-6+2E-6$$
$$= 6E-6(\text{per year})$$

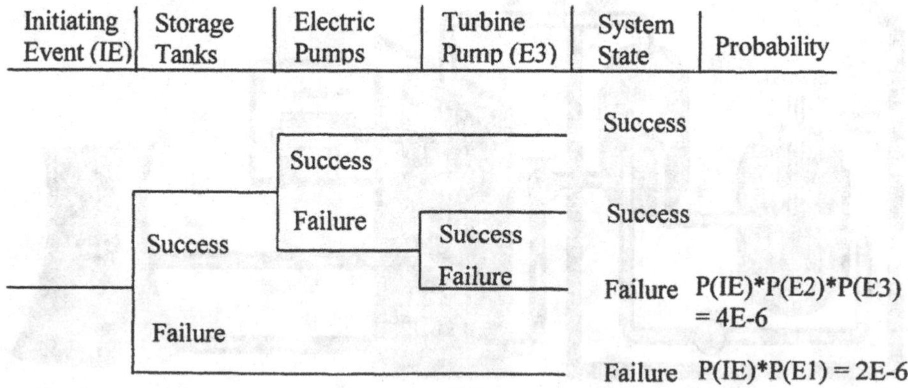

FIGURE 3.9 Event tree for loss of main feed water supply in a PWR.

3.6 A RISK EXAMPLE: THE TMI ACCIDENT

The most serious U.S. commercial reactor failure occurred on March 28, 1979, at the Three Mile Island (TMI) reactor near Harrisburg, PA. This was a PWR as shown in Figure 3.10. The reactor safety circuits correctly responded to a relatively minor initial fault and the sequence of shutting down the reactor began. However, it is worth highlighting some of the events in the sequence which were incorrectly interpreted by the operators causing a loss of cooling to the reactor core which resulted in partial melting of the core.

Human Error: The TMI-2 accident began when a water purifier on the secondary side blocked the coolant flow. A "stuck open" valve allowed water in the form of steam to escape from the reactor vessel. The emergency core cooling system operated as designed and provided makeup water for the core. The operators, believing that the pressurizer was "going solid" (being overfilled), shut off the emergency core cooling system. In reality, a steam bubble was forming in the core, giving the appearance that the system was going solid. The decay heat from the core boiled off the available water in the vessel, and without adequate cooling, the cladding and fuel began to melt. Before the operators resumed the flow of emergency coolant, a sizable portion of the core, about one-third to one-half, had melted (see Figure 3.11). The molten fuel and cladding dropped into the bottom of the vessel, which was full of water. This water was adequate to quench the molten material.

Radioactivity Release: A sizable amount of gaseous fission products escaped from the vessel through an open valve in the pressurizer. Most of the radioactive materials were confined to the containment building where it was prevented from being released into the environment. A small amount of radioactive material did escape. Some of the coolant water in the containment building was inadvertently pumped into the auxiliary building and some gas leaked into the environment. The releases were almost entirely noble gases, which are chemically inert and not

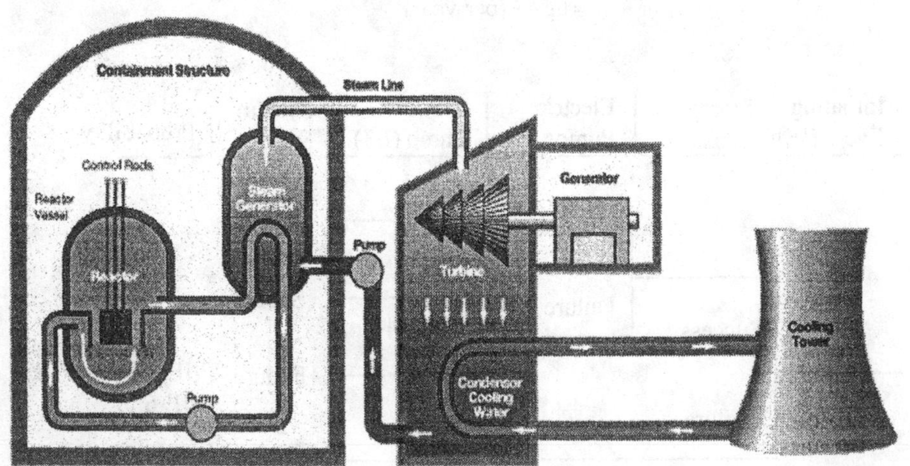

FIGURE 3.10 A Pressurized Water Reactor (PWR).

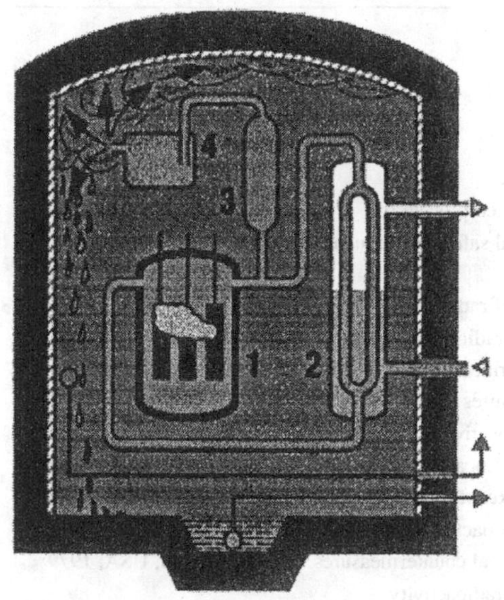

1. **Reactor**
2. **Once-through Vertical Steam Generator**
3. **Pressurizer**
4. **Quench Tanks or Pressurizer Relief Tank**

FIGURE 3.11 Partial core damage at Three Mile Island.

retained within the human body. Since the mid-1970s, event tree analyses have been used in various well-known risk assessment reports for radioactivity release from nuclear power reactors.

3.7 AN INTERNATIONAL RISK SCALE

When the media report an earthquake as over 7 on the Richter Scale, we know it's severe. One below 4 is not as severe and less than that may be quite minor. This is because the Richter Scale, named after its inventor Charles Richter, an American seismologist, is universally accepted as a measure of severity of earthquakes. It's actually a measure of the energy released by the earthquake. It runs from 1 to 10 and is logarithmic which means that an earthquake measuring 7 on the Richter Scale is ten times as severe as one measuring 6, in terms of ground movement.

There is an equivalent scale for conveying to the public the safety significance of events in the nuclear industry (Table 3.3). It is known as the International Nuclear Event Scale or INES. It was conceived by a group of experts from the International Atomic Energy Agency (IAEA) and the Nuclear Energy Agency of the Organization for Economic Co-operation and Development (OECD). It is only applicable to nuclear or radiological safety.

TMI was a major reactor accident (in financial cost) in that it was a partial melt-down of a modern power reactor core and it rated a level 5 on the INES. However, the radiological consequences of the accident were minimal. This was not simply a question of luck; it was due to the defense-in-depth safety principle used throughout the nuclear industry worldwide. The fuel was surrounded by

TABLE 3.3
International Nuclear Event Scale

Level	Descriptor	Criteria	Examples
0	Below Scale	No safety significance.	
1	Anomaly	Variation from permitted procedures.	
2	Incident	Incident with potential safety consequences on-site. Insignificant release of radioactivity off-site.	
3	Serious Incident	Very small release of radioactivity, a fraction of the prescribed limits. Local protective measures unlikely.	Vandellos, Spain, 1989. Possible acute health effects to a worker.
4	Accident Without Significant Off-Site Risks	Minor release of radioactivity of the order of prescribed limits. Local protective measures unlikely except for some food.	St-Laurent, France, 1980. Significant plant damage. Fatal exposure of a worker.
5	Accident with Off-Site Risks	Limited release of radioactivity. Partial implementation of local countermeasures.	Windscale, UK, 1957. Three Mile Island, USA, 1979
6	Serious Accidents	Significant release of radioactivity. Full implementation of local countermeasures.	
7	Major Accident	Major release of radioactivity. Widespread health and environment effects.	Chernobyl, USSR, 1986

three layers of containment: the fuel "clad", the reactor pressure vessel, and the containment building. Only the first failed, due to overheating, although the pressure vessel containment was temporarily lost when the pressure relief devices operated. This design prevented an uncontrolled release of radioactivity into the environment.

Defense-in-depth is a very important safety principle. Its application to design for risk engineering will be further discussed in the next chapter. Figure 3.12 illustrates the defense-in-depth safety concept, with bold path reflecting TMI incident.

Risk assessments are widely used by agencies of the federal government that make safety, health, or design decisions about risk-posing facilities or equipment. Examples of U.S. government agencies that use engineering risk assessments are as follows:

- Department of Energy, in evaluating the radiological and chemical risks from various types of nuclear and non-nuclear facilities.
- Department of the Interior, in analyzing dam safety, assessing damage from ecological disasters, and helping to predict natural hazards, such as earthquakes, floods, or volcanoes. Much of this risk assessment work is directed toward improving the probability distributions that describe the recurrence of these natural hazards and their possible intensity.
- Federal Aviation Administration, in analyzing potential collision scenarios, such as the simultaneous approach of two aircraft on closely spaced, parallel runways in inclement weather.

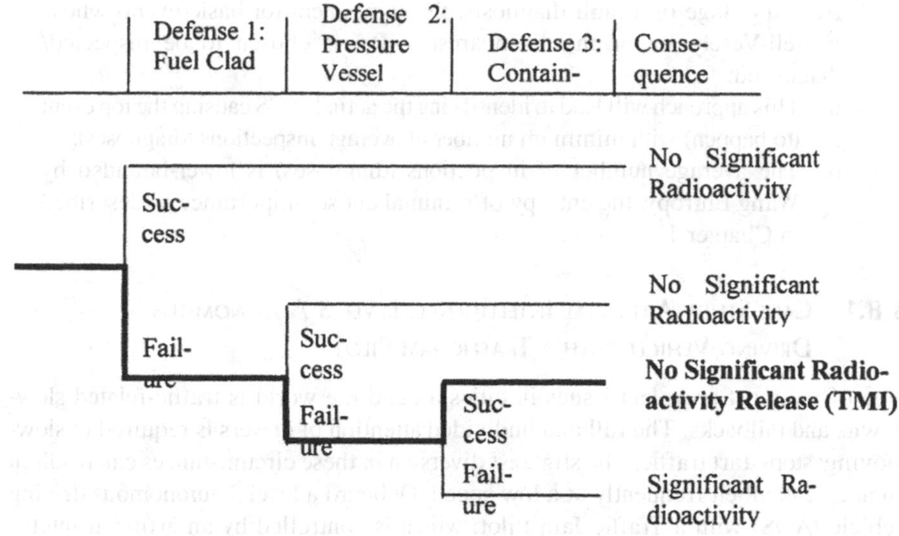

FIGURE 3.12 An event tree illustrating the defense-in-depth safety principle (bold path reflects TMI incident).

- National Aeronautics and Space Administration, in assessing the possibility of shuttle accidents that might result in the release of radioactive material from radioactive power sources.
- Nuclear Regulatory Commission, in analyzing risks of low-level radioactive waste disposal, evaluating performance of high-level waste repositories, and evaluating risks associated with nuclear power plant accidents.

3.8 INFORMATION GAIN: CAUSAL RISK ASSESSMENT BASED ON WANG ENTROPY

If the original cut set importances are q_1, q_2, •••, q_n. After a set of inspections, the cut set importances are updated to be I_1, I_2, •••, I_n. The information gain obtained from the set of inspections can be evaluated by

$$\sum_{i=1}^{n} I_i log_2 \frac{I_i}{q_i} = \sum_{i=1}^{n} I_i - \sum_{i=1}^{n} q_i = W_H - \sum_{i=1}^{n} I_i log_2 q_i = W_H - \sum_{i=1}^{n} I_i log_2 q_i$$

which is called the relative entropy of cut set importances I_1,$I2$, •••, I_n with regard to the original distribution $q1$, $q2$, •••, q_n. W_h denotes Wang Entropy, a measure of system complexity as described in Chapter 1. The steps for performing causal risk assessment based on Wang Entropy are as follows:

1. Develop quantitative FTA/calculate importance measures;
   a. Generate minimal cut sets (MCSs)
   b. Generate cut set importance
   c. Generate Fussell-Vesely importance (for basic events)

2. At every stage of a fault diagnosis, the component (or basic event) whose Fussell-Vesely importance is nearest to 0.5 is chosen to be inspected/diagnosed;
   a. This approach will lead to identifying the actual MCS causing the top event (to happen) with minimum number of average inspections (diagnoses).
   b. The average number of inspections (diagnoses) is lower-bounded by Wang Entropy, the entropy of minimal cut set importance as described in Chapter 1.

3.8.1 Case Study Artificial Intelligence: Level 3 Autonomous Driving Vehicle with a Traffic Jam Pilot

One of the most prevalent issues in cities around the world is traffic-related slowdowns and tailbacks. The full and undivided attention of drivers is required in slow-moving, stop-start traffic. The slightest diversion in these circumstances can result in an accident, albeit frequently at a low speed. Onboard a level 3 autonomous driving vehicle (ADS) with a Traffic Jam Pilot, which is controlled by an artificial intelligence algorithm and achieves level 3 autonomous driving in heavy traffic at speeds of up to 60 km/h, this issue becomes much less challenging.

Both machine learning and artificial intelligence are burgeoning fields with a lot of potential. Piloted vehicles will eventually be able to respond effectively in extremely complicated situations, just like a human driver would, possibly even better, thanks to artificial intelligence. Artificial intelligence is an area of information technology that focuses on giving machines human-like skills. Using machine learning, for instance, might make this possible.

Therefore, machine learning is a prerequisite for artificial intelligence. Mathematics and statistics serve as the foundation for this. Algorithms will independently discover patterns and rules in even the most complex situations, and they will base their choices on them. Recent years have seen significant advancements in the study of artificial neural networks, which mimic the signal connections found in the human brain. Deep learning simulates the brain's neural networks on a computer. A large data set and powerful processing resources are needed for this.

3.8.1.1 Risk Engineering Goal for a Traffic Jam Pilot

The highest (top) level risk engineering requirements for each item are called risk engineering goals. To ascertain the risk engineering requirements for the system being designed, the process includes risk assessments. To determine the fundamental requirements or risk engineering goals for the function and any linked systems, a risk assessment needs to be performed.

The following risk engineering goal is one of the results of the risk assessment for the Traffic Jam Pilot:

- "Avoid insufficient vehicle deceleration when Traffic Jam Pilot is operational".

In essence, the car must be driven to an "operating mode without an unacceptable level of risk" in order to achieve a safe state in the event of potentially damaging circumstances (e.g., the driver takes over the control of the vehicle or the autonomous driving functionality switched to a degraded operation mode and finally into a safe state, for example, a decelerated trend). Other risk engineering goals include the following:

- Operating mode changes that are unauthorized or unintended must be avoided.
- Safe Halt operating mode change must be verified to be intended and authorized.
- Over-actuation of the steering must be avoided.
- Unintentional activation of the anti-lock brakes must be avoided.
- Acceleration that wasn't meant must be avoided.
- The ability to detect driver intervention must be guaranteed.
- The Human Machine Interface (HMI) must ensure that the true operating mode is displayed.

In order to prioritize strategies and allocate scarce resources, risk assessments are used.

3.8.1.2 Risk Management for a Traffic Jam Pilot

For risk management, the risk assessment for other items near the ego vehicle, the vehicle that houses the sensors used to understand its surroundings, is carried out by the risk manager (RM). This module receives data from sensors in the environment perception block, such as light detection and ranging (LiDAR), radar, stereo cameras, and sensor fusion algorithms. Additionally, RM receives information about the vehicle's dynamics, including the speed, acceleration via the inertial measurement unit (IMU), steering wheel angle, gear, brake, throttle, etc. The sensor data is analyzed and categorized. The module conducts categorization and feature extraction from the objects.

3.8.1.3 AI Was Unable to Identify Images for Traffic Jam Pilots

The "Traffic Jam Pilot" system, when used by the driver, monitors traffic conditions on the highway and, if it deems it safe to do so, automatically assumes control of the vehicle. It lets the car to follow the one in front of it and maintain its lane position without any driver input.

This is a common instance of an AI failure potentially for Traffic Jam Pilots. The breakthrough in picture recognition, often known as computer vision, launched the triumphal march of deep learning, a group of techniques frequently used to build AI, about 20 years ago. Before going on to more challenging and demanding tasks, it solved previously impossible tasks like differentiating between people and others. The notion that computer vision is a reliable, dependable, and unlikely to fail technology is now universally accepted. Seven thousand five hundred unedited nature photos were acquired by researchers from Berkeley, Chicago, and the University of Washington. These photos baffled even the most advanced computer vision algorithms. Even the most tested and reliable algorithms can occasionally go wrong.

3.8.1.4 Health Monitoring System for Next-Generation Autonomous Driving Using Machine Learning

The development of various integrated smart sensors has made it possible to monitor the health of such sophisticated ADSs online. It is possible to continuously monitor the state of the various ADS components and take the required precautions before they break down. This case study, which is illustrated in Figure 3.13, explains how to use various embedded smart sensors to detect and monitor the health issues or failure of various novel Automatic Emergency Braking (AEB) system, subsystems, and components of a self-driving car. It also explains how to apply Wang Entropy as a measure of the difficulty of system diagnosis to enable proactive maintenance to prevent breakdowns.

Example 3.7

For a novel Automatic Emergency Braking system of a self-driving car, a control system injecting liquid from tanks g, h, and i to a mixer is shown in Figure 3.13. After receiving the electrical signal, sensor b will open the two isolation valves c and d. The two pumps e and f will be started on demand. A gravity-driven loop a, which functions when the pump-driven loop fails, is also provided.

With Wang Entropy as a measure of the difficulty of system diagnosis, along with decision-tree-based machine learning, the following three methods have also been used for the novel AEB system of a self-driving car being used as the case study:

1. one for system monitoring,
2. one for time series forecasting, and
3. one that combines the other two methods into a single, comprehensive monitoring and prediction process.

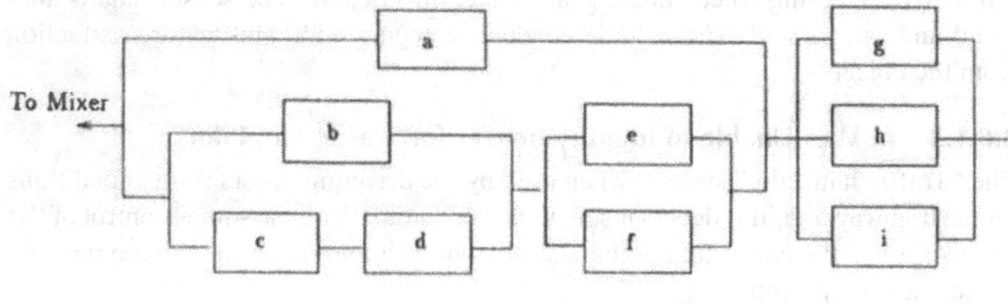

a – a gravity-driven loop; b – sensor;

c, d – isolation valves; e, f – pumps;

g, h, i — tanks.

FIGURE 3.13 Case study of Wang Entropy as a measure of the difficulty of system diagnosis for a novel AEB system of a self-driving car.

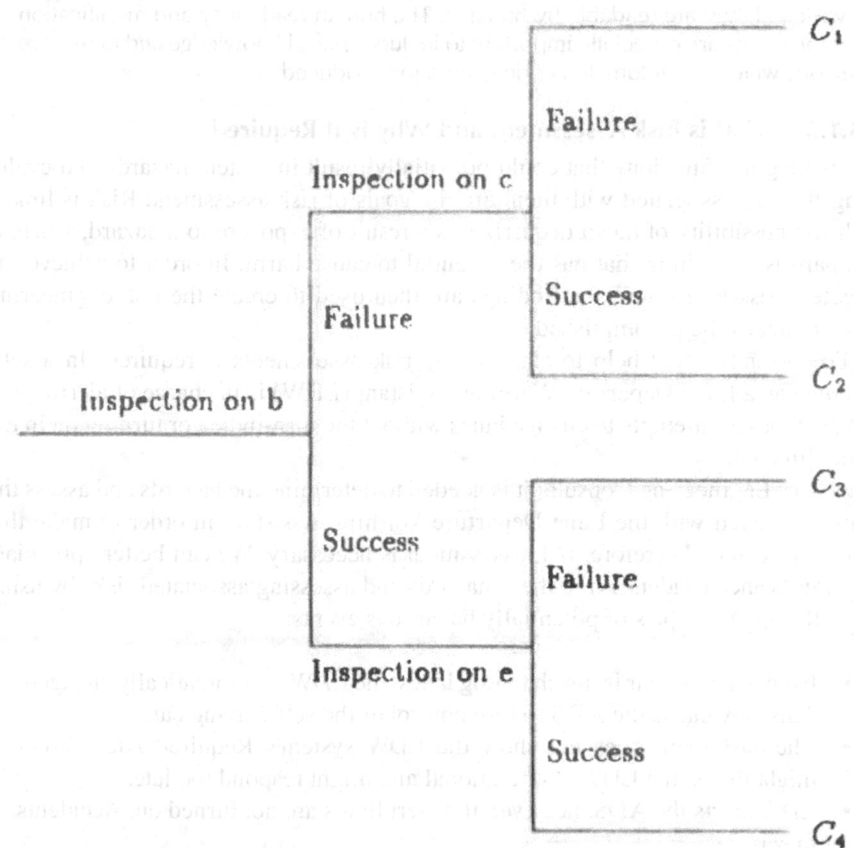

FIGURE 3.14 Case study of optimizing the inspection sequence based on decision tree based machine learning for the health monitoring system in the automatic emergency braking system of a self-driving car.

The combination of the three techniques enables the prediction of potential failures. The AEB system of a self-driving car uses decision-tree-based machine learning to optimize the inspection sequence. This is shown in Figure 3.14. The advantage of decision trees is that the resulting code can be processed quickly and simply, can be easily modified by human experts, and can be read by humans. It is also crucial to make the outputs readable and modifiable by humans in order to include specialized knowledge and eliminate faults that the automated code generation caused.

Together the three methods allow the forecasting of possible failures. Figure 3.14 illustrates optimizing the inspection sequence based on Wang Entropy (see Chapter 1 for details) and decision-tree-based machine learning for the Health Monitoring System in the AEB system of a self-driving car. Wang Entropy enables to find the fault(s) causing system failure averagely with the least number of inspections. Decision trees have the advantage that the generated code can be fast and easily processed, they can be altered by human experts without much work and they are readable by humans. The human readability and modification of the results are especially important to include special knowledge and to remove errors, which the automated code generation produced.

3.8.1.5 What is Risk Assessment and Why Is It Required?

Identifying malfunctions that could potentially result in system hazards and evaluating the risk associated with them are the goals of risk assessment. Risk is linked with the possibility of harm occurring as a result of exposure to a hazard, whereas a hazard is something that has the potential to cause harm. In order to achieve the targeted "risk-free state", the findings are then used to create the risk engineering goals that must be accomplished.

This example will help to clarify why risk assessments is required. In a self-driving car, a Lane Departure Warning Assistant (LDW) is in charge of alerting the driver if the car attempts to change lanes without the turn-indicator turning on in the same direction.

A Risk Engineering Consultant is needed to determine the hazards and assess the risks associated with the Lane Departure Warning Assistant in order to make this feature risk-free. Therefore, risk assessment is necessary. We can better appreciate the significance of identifying these hazards and assessing associated risks by using the following examples of potentially hazardous events:

- Even when the car is not changing lanes, the LDW is automatically engaged. This may cause the ADS to lose control of the self-driving car.
- The dashboard does not show the LDW system's Required Alert. ADS might think that LDW is operational and might respond too late.
- LDW alerts the ADS; however, the alert lights are not turned on. Accidents may occur.

3.8.1.6 Using the Hazard and Operability (HAZOP) Technique to Identify Malfunctions

3.8.1.6.1 What is HAZOP?

A Hazard and Operability (HAZOP) study is a planned or existing process or operation's structured and methodical assessment in order to identify and assess problems that might provide risks to people or equipment or obstruct effective operation.

3.8.1.6.2 Is a HAZOP a Risk Assessment?

Risk assessment teams often employ the systematic technique known as a HAZOP analysis, or HAZOP, to find potential risks and problems with a system or process's operability.

HAZOP is a technique that is frequently used for assessing potential risks in a system and locating operability problems, which are prone to produce non-conforming products.

3.8.1.6.3 Controllability of Autonomous Driving System (ADS)

When a safety goal is violated due to a failure or malfunction of any automotive component, controllability (C) assesses the extent to which the ADS can maintain control of the vehicle.

3.8.1.6.4 HAZOP Example: Automatic Emergency Braking (AEB) System of a Self-Driving Car

- None: There is no braking command generated by EB.
- Less: EB applies less braking than the maximum.
- Late: EB applies maximum braking, but with a 2-second delay.
- Reverse: Instead of braking, EB generates an accelerating command.
- Before: EB applies maximum braking prior to the detection of a potential collision.

Using the HAZOP technique to identify malfunctions helps to develop intermediate events and basic events in FTA; see Section 3.4 for risk assessment.

BIBLIOGRAPHY

Brombacher, A. C. (1992), *Reliability by Design*, Wiley & Sons, Hoboken, NJ.

Center for Chemical Process Safety. (1989), *Process Equipment Reliability Data*, American Institute of Chemical Engineering, New York, NY.

Gillett, J. (1996), *Hazard Study & Risk Assessment: A Complete Guide*, Interpharm Press, Incorporated, New York, NY.

Kaplan, S. and Garrick, B. J. (1981), "On the Quantitative Definition of Risk," *Risk Analysis*, Vol. 1, No. 1, pp. 11–27.

Kumamoto, H. and Henley, E. J. (1995), *Probabilistic Risk Assessment & Management for Engineers & Scientists*, IEEE Press, New York, NY.

Lemons, J. (1995), *Risk Assessment (Readings for the Environment Professional)*, Blackwell Scientific, Piscataway, NJ.

Marsili, G., Vollono, C., and Zapponi, G. A. (1992), "Risk Communication in Preventing and Mitigating Consequences of Major Chemical Industrial Hazards," *Proceedings of the 7th International Symposium on Loss Prevention and Safety Promotion in the Process Industries*, Taormina, May 4–8, Vol. 168, pp. 1–13.

Melchers, R. E. and Stewart, M. G. (1994), *Probabilistic Risk & Hazard Assessment*, Balkema (A. A.) Publishers, Netherlands.

Melchers, R. E. and Stewart, M. G. (1995), *Integrated Risk Assessment: Current Practice and New Directions: Proceedings of the Conference on Integrated Risk Assessment*, Ashgate Publishing Company, Newcastle, NSW, Australia.

Roush, M. L., Weiss, D., and Wang, J. X. (1995), "Reliability Engineering and Its Relationship to Life Extension," An invited paper presented at the 49th Machinery Failure Prevention Technology Conference, Virginia Beach, VA, April 18–20.

Society for Risk Analysis Europe. (1996), "Risk in a Modern Society Lessons from Europe," The 1996 Annual Meeting for the Society for Risk Analysis - Europe, University of Surrey, Guildford.

Vose, D. (1996), *Risk Analysis: A Quantitative Guide to Monte Carlo Simulation Modelling*, Wiley Liss Inc., Hoboken, NJ.

Wang, J. X. (1996), "Complexity as a Measure of the Difficulty of System Diagnosis," *International Journal of General Systems*, Vol. 24, No. 3, pp. 257–269.

Wang, J. X. (2017), *Industrial Design Engineering: Inventive Problem Solving*, CRC Press, Boca Raton, FL.

Wang, J. X. (2019a), "Complexity as a Measure of the Difficulty of System Diagnosis in Next Generation Aircraft Health Monitoring System," SAE Technical Paper 2019-01-1357, doi:10.4271/2019-01-1357.

Wang, J. X. (2019b), "A Dynamic Fault Tree Approach for Time-Dependent Logical Modeling of Autonomous Flight Systems," SAE Technical Paper 2019-01-1358, doi:10.4271/2019-01-1358.

4 Design for Risk Engineering

The Art of War Against Failures

4.1 CHALLENGER: CHALLENGING ENGINEERING DESIGN

On January 28, 1986, the space shuttle Challenger exploded during launch, killing the crew and virtually stopping U.S. exploration of space for two years. The presidential commission formed to study the disaster concluded in its report that the principal cause of the explosion was "a failure in the joint between the two lower segments of the right Solid Rocket Motor", and specifically "the destruction of the seals that are intended to prevent hot gases from leaking through the joint during the propellant bum…". The commission attributed the destruction of these seals to the failure of an O-ring seal in the aft field joint of the right-hand solid booster. In fact, the commission blamed the failure on the design of the joint:

> The Space Shuttle's Solid Rocket Booster problem began with the faulty design of its joint and increased as both NASA and contractor management first failed to recognize it as a problem, then failed to fix it and finally treated it as an acceptable flight risk.

Figure 4.1 shows the major parts of the solid-fuel boosters.

The booster that failed is one of two solid-fuel boosters designed to help the shuttle reach the velocity needed for orbit (see Figure 4.1). The booster is shipped to the launch site in sections and is assembled onsite. The aft field joint is one of the joints made during the final field assembly. The solid propellant, cast inside the booster, bums from the center outward. The burning fuel expands, causing great pressure inside the booster and giving the booster its thrust. Essentially, the booster is a thin-walled, large-diameter pressure vessel. As the fuel is burned, the temperature of the outer wall and the joint also increases.

As shown in Figure 4.2, the solid rocket booster segments are joined by means of a tang-and-clevis arrangement, with two O-rings to seal the joint and 177 steel pins around the circumference to hold the joint together. The clevis is the U-shaped section of the lower segment into which the *tang* from the upper segment fits. The zinc-chromate putty acts as an insulation that under pressure, according to the design, would behave plastically and move toward the O-rings. This would pressurize the air in the gap between the two segments, which would, in turn, cause the first O-ring to extrude into the gap between the clevis and the tang, sealing the joint. If the first O-ring failed, the second one could take the pressure. For this scenario, ideally hot gases are shielded from the joint by the zinc-chromate putty. For the scenario shown on the right, the high pressure creates a blowhole in the putty, allowing the O-rings to move into the positions

DOI: 10.1201/9781003371014-4

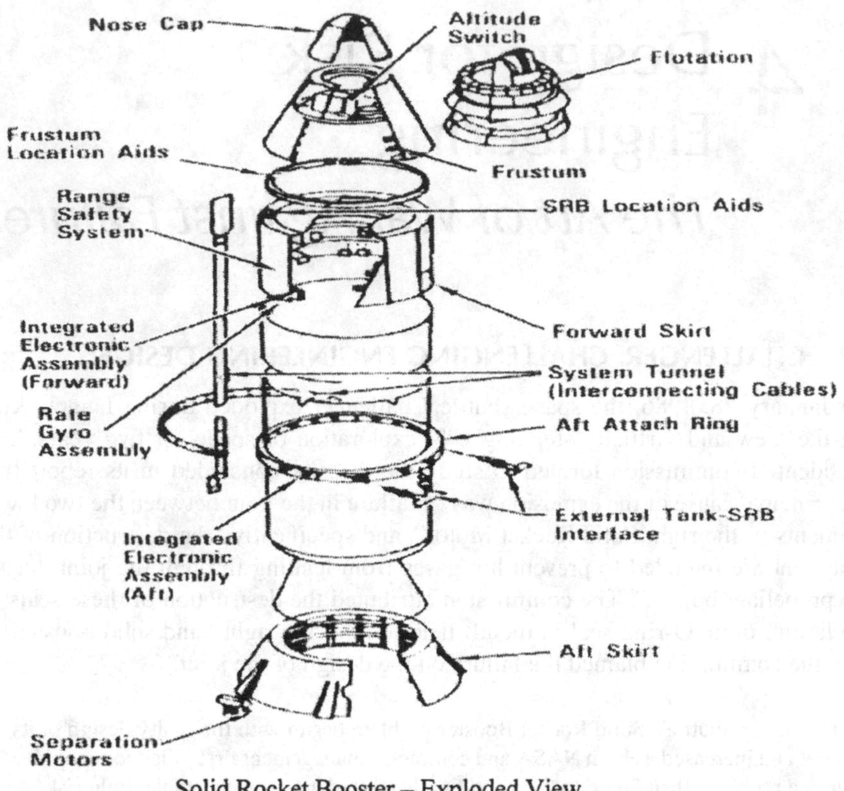

FIGURE 4.1 Major parts of the solid rocket booster.

needed to seal the joint as the gap between tang and clevis expands. Through the blowhole, however, gases penetrate and in passing the O-rings wear away the O-rings.

This design was based on similar joints designed for the very reliable Titan-III rocket. However, the tang on the Challenger's booster joint was longer and flexed under pressure more than that on the Titan. Early experimental tests of the shuttle's joint showed that the flexing of the joint caused the tang to rotate in the clevis, opening the joint at the O-rings. This condition was more serious when the O-rings were cold, as they lost resiliency and could not change from their compressed shape, as shown in Figure 4.2, to the shape needed to fill the opening.

Let's review the design process that resulted in this joint. We can find the origin of the trouble in the engineering specification development phase, where the goal is to understand the customer's requirement for risk management and translate these requirements into specifications for engineering design. The presidential commission's report makes evident that this phase of the design process left much to be desired. In fact, the very first recommendation from the commission specified that *"the joints should be fully understood, tested and verified"*. As illustrated later in Section 4.2, the understanding about customer need, "what", and the design specification, "how", is the key to developing a successful engineering project.

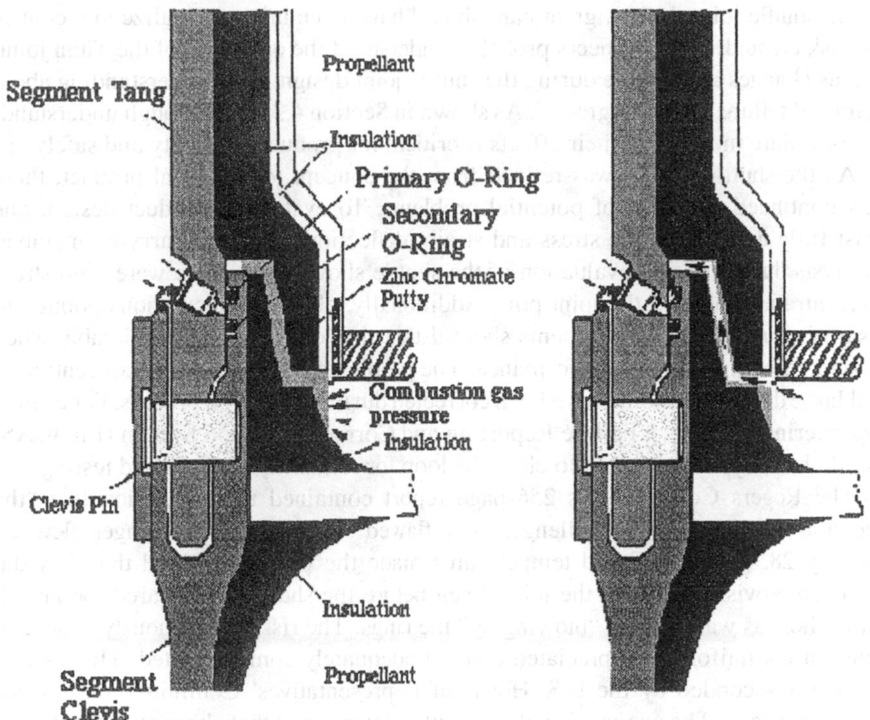

FIGURE 4.2 Cross-section of the booster field joint.

Three competing factors affect the design process: cost, quality, and time. It appears from the commission's report that low cost eventually drove the original shuttle contract. The adopted proposal said that the designers need only to modify an existing Titan-quality design. Time pressures, as well as cost, kept the joint from being totally redesigned. When it checked into the history and performance of the O-ring sealing system, the Rogers Commission members were amazed to find that the O-rings had failed regularly, if only partially, on previous shuttle flights. Although both NASA and Thiokol were concerned about the frailty of these seals, they chose to forgo a time-consuming redesign of the system. Both had come to regard the observed O-ring erosion as an acceptable risk because the seal had only *partially* failed before.

In the second phase of the design process, the conceptual development phase, the shuttle booster concept was taken from the Titan rocket with relatively little modification. Although it is a good design practice to utilize past concepts, great care must be taken to ensure that the past concept is really applicable to the new use environment. The differences between the shuttle and the Titan made the existing joint design questionable. The Challenger booster's O-rings took the pressure of combustion, whereas the single O-ring in the Titan did not. In the Titan, the insulation was tight-fitting and the O-ring had only to take the pressure of any leakage through the insulation. Since the critical functions of load carrying and sealing were combined in the shuttle joint design, great care should have been taken to realize and control the risk created. The engineers probably understood the operation of the Titan joint, but, as changes were made during the shuttle joint design, their understanding about potential failure modes degraded. As shown in Section 4.3, the thorough understanding of failure modes and their effects is critical for product reliability and safety.

As the shuttle design was refined from the concept to an actual product, there was continued evidence of potential problems. To evaluate a product design, one must fully understand the stress and strain (deflection) in a load-carrying member. Analysis during design evaluation of the shuttle showed that there were high stress concentrations around the joint pins. Additionally, although calculations pointed to a satisfactory design, experiments showed that the joint rotated considerably when loaded, affecting the seal performance. The problems with high stress concentration and large deflection should have been corrected during the design process. For critical engineering projects, a Failure Reporting and Corrective Action System (FRACAS) should be fully implemented to close the loop for reliability analysis and testing.

The Rogers Commission's 256-page report contained the conclusion that "the decision to launch the Challenger was flawed". When the Challenger flew on January 28, 1986, the frigid temperatures made the O-rings so hard that they did not even provisionally seal the joint. Even before the shuttle had cleared the launch tower, hot gas was already "blowing by" the rings. The risk was obviously unacceptable but not uniformly appreciated and not adequately communicated. This assessment was seconded by the U.S. House of Representatives' Committee on Science and Technology. The congressional committee determined that although the technical problem had been recognized early enough to prevent the disaster, "meeting flight schedules and cutting cost were given a higher priority than flight safety". In Chapter 9, we will address the impact of administration on risk engineering and management.

The Challenger did challenge the traditional engineering design process and this example illustrates the need for more rigorous design within a proper risk engineering process.

4.2 GOAL TREE: UNDERSTAND "WHAT" AND "HOW"

Failure is the enemy of engineers. A universal concept unifying all engineering fields is the art of war against failures. Failure can take non-technical forms. A design can be considered a failure if it is environmentally unsound or aesthetically unsatisfying. Such criteria should be taken into account from the beginning of the design phase, just as the strength of materials. Engineering also has to a lot to do with economics. A too-expensive product will fail to survive the competition, a non-technical failure mode that can cause the failure of engineering projects.

A famous design story centers on the Mariner IV satellite project (Figure 4.3a). The satellite was to be packaged in a rocket, its solar panels folded against its sides. After launch, the satellite was to spin so that the solar panels would unfold by centrifugal force and be locked in a straight-out position. Because these panels were quite large and very fragile, there was concern that they would be damaged when they hit the stops that determined their final position. To address this problem, the major aerospace firm that had the Mariner contract initiated a design project to

FIGURE 4.3a Mariner IV satellite.

develop a retarder (dampener) to gently slow the motion of the panels as they reached their final position.

The constraints on the retarders were quite demanding: they would have to work in the vacuum and cold of space, work with great reliability, and not leak since any foreign substance on the panels would harm their capability of capturing the sun's energy during the satellite's nine-month mission to Mars. Millions of dollars and thousands of hours were spent to design these retarders; yet after extensive design work, testing, and simulation, no acceptable devices evolved. With time running out, the design team ran a computer simulation of what would happen if the retarders failed completely; to the team's amazement, the simulation showed that the panels would be safely deployed without any dampening at all. In the end, they realized that there was no need for retarders, and Mariner IV successfully went to Mars without them.

This story is an example of how a lot of time and money can be wasted designing the wrong product. Surveys show that poor product definition is a factor in 80% of all time-to-market delays. Understanding what to do and how to do it may seem a simple task; unfortunately, it often is not. Sometimes doing product development and planning is like trying to get your arms around a 500-pound marshmallow.

One possible solution is the use of a goal tree. A goal tree is a success-oriented logic structure that can be used to organize complex systems and their engineering knowledge into a format suitable for problem solving. Defining "what is the customer need" is a common but difficult task because true "customer needs" are often disguised in a variety of ways. It takes a skillful individual to analyze a situation and extract the real "what" from a sea of information. Ill-defined or poorly posed "what" questions can lead novice (and not so novice) engineers down the wrong path to a series of impossible or non-useful solutions. Defining the "real need" is critical to finding a good solution. The following example provides evidence of how millions of dollars and thousands of hours can be wasted by poor problem definition (what) and solution (how).

Example 4.1

Flow-meters, such as the ones at gasoline pumps to measure the number of gallons of gas delivered to your gas tank, are commonplace in industry. A flow-meter was installed in a chemical plant to measure the flow rate of a corrosive fluid. A few months after installation, the corrosive fluid had eaten through the flow-meter and began to leak onto the plant floor. The instructions given to solve the perceived problem: *"Find material from which to make a flowmeter that will not corrode and cause leakage of the dangerous fluid"*. An extensive, time-consuming search was carried out to find such a material and a company that would construct a cost-effective flowmeter. None was found. However, the **real problem** was to prevent the flow-meter from leaking. The ultimate solution was to institute a program of simply replacing the existing flow-meters on a regular basis before corrosion caused a failure.

As shown in Figure 4.3b, for the meter system described above, a goal tree has been developed by first defining a goal based on the customer need, i.e., "provide accurate flow readings without peripheral damage", and then identifying sub-goals which must necessarily be achieved if the goal is to be achieved. The sub-goals include:

- Prevent the flow-meter from leaking; and
- Provide accurate flow measurements.

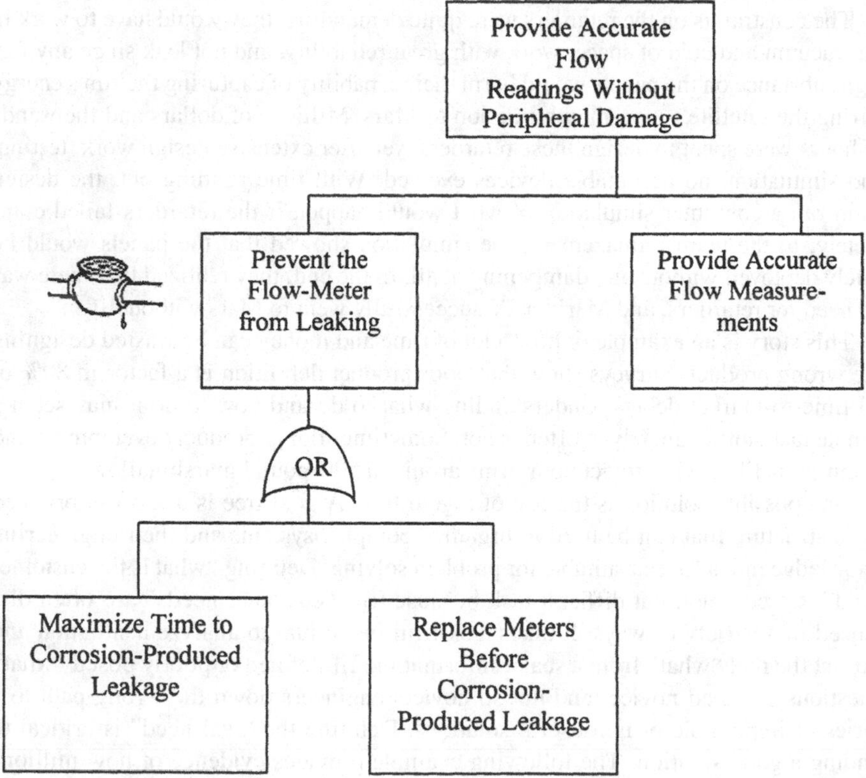

FIGURE 4.3b A goal tree for a meter system (Example 4.1).

The goal tree provides a structured process for clearly specifying the customer's needs and desires. The procedure translates needs and desires into requirements at each stage of product or service design, development, execution, and delivery. The tree development is a method which takes into account customer requirements at each stage of product or process development. As a formal process for understanding, recording, and quantifying the interactions between the various elements of a product or service, goal tree analysis proves its worth for managing engineering risk.

The power of goal tree analysis lies in the fact that, done with care and integrity, it lays bare an organization's processes and how these processes interact to create customer satisfaction and profit:

- Start with Customer Requirements – "provide accurate flow readings without peripheral damage" in Example 4.1;
- Translate to Design Requirements:
  1. "Precision" – Provide accurate flow measurements;
  2. "Meter Reliability" – Maximize time to corrosion-produced leakage;
  3. "Preventive Maintenance"- Replace meters before corrosion-produced leakage;
- Translate to Part Characteristics;
- Translate to Purchasing and Process Plans;
- Translate to Specific Operations, Conditions, or Controls.

Design is an activity that recognizes the goals or purposes of products or systems. It is an activity that shapes its objects – creates their form – in accordance with the goals or purposes of those objects. The product design process is a process of function allocation that identifies product purposes – such as functions – and allocates them to a structural product. The successive decomposition of identified goals continues until the goal resolution reaches a level of knowledge for evaluating engineering specifications and alternatives.

With the rolling out of the 777 model, Boeing has unleashed the aircraft of the future. The 777 has incorporated revolutionary changes in design philosophy. Among its most important changes is the incorporation of sophisticated computer flight control systems. These systems have changed the way pilots will fly the next generation of Boeing aircraft. The new fly-by-wire system in the 777, the first ever in Boeing's fleet of aircraft, gives the pilot much more information and feedback concerning the operation of the plane, than did the older electromechanical system in the older Boeing models. The top-level objective is to "improve reliability, performance and operability of the airplane system".

New fly-by-wire systems have antiquated the older electromechanical systems present in models such as the 747, the 737, and ones preceding those. The design requirements of the new flight control system are flowed down by the following sub-goals.

Provide More Precise Information/Feedback to Pilots: The older system did not provide the pilot much feedback information such as possible equipment malfunctions, exact airplane heading, or local weather conditions.

The 777 has two 8×8-inch flat panel, LCD screens for displaying accurate information from the flight computers for the pilots. The two displays are for primary flight information and for navigation. These have replaced the older system of gyroscopic dials and analog indicators which used to clutter the cockpit. In addition, there is a multifunction display in the center console to give the pilots additional flight information such as speed and altitude. All these displays are of high resolution and enable the pilots to read even the minute details clearly.

The 777 also incorporates a data link capability that allows the pilot to receive weather and other flight information. This data can be displayed on the screen or printed on a printer. The data link feature has been modified to receive air traffic control transmissions on speed, heading, and altitude assignments. Satellite communications and a global positioning system give the aircraft additional communication and navigational capability.

Reduce Weight: In older systems, much of the control was through mechanical devices. Steel cables ran from the cockpit to the wings and tail in order to move control surfaces. The older system was much heavier, due to the added weight of the steel cables and other mechanical devices.

Instead of the older system of steel cables, there are computer-controlled electrical systems that operate through thin wires, controlling the movements of the rudder, the elevators, and the ailerons. This system reduces weight and insures an extremely precise flight adjustment.

Improve Reliability: Since most of the control actuators in the older system were mechanical devices subject to wear and fatigue, *the possibility was present that* there could be a mechanical failure either in-flight or on the ground. Such mechanical defects are hard to detect without the proper feedback devices.

One of the most important aspects of the new flight control system is the presence of various safety features that are built-in. The system incorporates numerous backup features so that if one unit were to fail, another would take over. The computers on-board have stall protection features that virtually prevent a stall, notwithstanding an intentional stall by the pilot. There are other features that automatically compensate for engine failure, a yaw damper which stabilizes the horizontal roll during turbulent flying conditions, and over-speed protection. There are backup electrical systems which consist of backup generators and a permanent magnet generator driven by the engine that is to be used in case of total electrical failure.

Enhance Maintainability: Maintenance routines and checking procedures were time-consuming with the numerous mechanical devices present in the aircraft.

The new instrumentation panel displays aircraft altitude, navigation information, and engine data right on the LCD displays. If a problem develops, the screen displays the malfunctioning part and a list of ways to correct the problem. If the problem involves control over a moving surface, that item is displayed on the screen, telling the pilot exactly what is wrong and the devices he can use with. The older aircraft left much room for human error as most of the control was done by the pilots.

As shown in Figure 4.4, a goal tree can be structured with rigorous hierarchy by applying two rules for each goal or sub-goal: (a) looking directly above should reveal "why" any specific goal or sub-goal must be achieved; (b) looking downward to the next decomposition should reveal "how" the specific goal or sub-goal is satisfied. The goal tree developed by careful adherence to these rules will be rigorously hierarchical and therefore extremely useful because of the embedded cause-consequence relationships.

The 777 is the result of a total change in the design policy of Boeing. The new fly-by-wire flight control system is a reflection of that change. Through the use of sophisticated computers and displays, the new system enables the pilot to get more information about the aircraft, fly accurate flight paths, and ensure the safety of the aircraft and the passengers. This type of flight control system has worked extremely well in the Space Shuttle and in the Airbus series of aircraft.

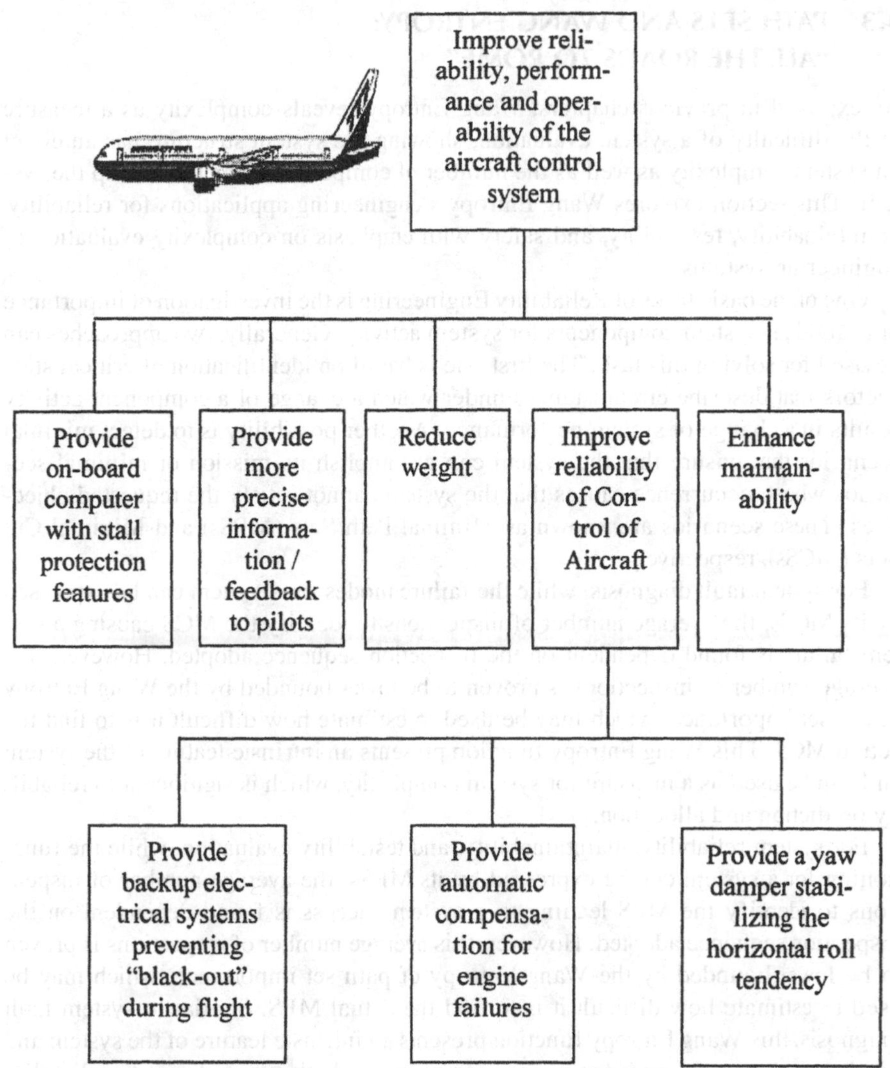

FIGURE 4.4 A goal tree illustrating the new Boeing 777's flight control system.

To design the 777, Boeing organized its workers into 238 cross-functional "design build teams" responsible for specific products. The teams used 2,200 terminals and a computer-aided three-dimensional interactive application (CATIA) system to produce a "paperless" design that allowed engineers to simulate assembly of the 777. The system worked so well that only a nose mockup (to check critical wiring) was built before assembly of the first flight vehicle which was only 0.03 mm out of alignment when the port wing was attached. Boeing also included customers and operators, down to line mechanics, to help tell them how to design the plane.

For large complex engineering projects, a goal tree is a valuable tool for identifying and specifying functional requirements before undertaking the detailed structural design solution.

4.3 PATH SETS AND WANG ENTROPY: "ALL THE ROADS TO ROME"

As explored in previous chapters, Wang Entropy reveals complexity as a measure of the difficulty of a system evaluation, showing the system structure has an effect on system complexity as well as the number of components which make up the system. This section explores Wang Entropy's engineering applications for reliability, maintainability, test ability, and safety with emphasis on complexity evaluation of engineering systems.

One of the basic tasks of Reliability Engineering is the investigation of importance of individual system components for system activity. Generally, two approaches can be used for solving this task. The first one is based on identification of critical state vectors that describe circumstances under which a change of a component activity results in a change of system performance. Another possibility is to detect minimal scenarios that ensure that the system can accomplish its mission or minimal scenarios whose occurrence causes that the system cannot satisfy the requested objectives. These scenarios are known as Minimal Path Sets (MPSs) and Minimal Cut Sets (MCSs), respectively.

For system fault diagnosis, while the failure modes for a system can be expressed by its MCSs, the average number of inspections to identify the MCS causing a system failure is found dependent on the inspection sequence adopted. However, this average number of inspections is proven to be lower-bounded by the Wang Entropy of cut set importance, which may be used to estimate how difficult it is to find the actual MCS. This Wang Entropy function presents an intrinsic feature of the system and can be used as a measure for system complexity, which is significant to reliability prediction and allocation.

For system reliability, maintainability, and testability evaluation, while the functioning for a system can be expressed by its MPSs, the average number of inspections to identify the MPS leading to a system success is found dependent on the inspection sequence adopted. However, this average number of inspections is proven to be lower-bounded by the Wang Entropy of path set importance, which may be used to estimate how difficult it is to find the actual MPS. Similar to system fault diagnosis, this Wang Entropy function presents an intrinsic feature of the system and can be used as a measure for system complexity, which is significant to reliability prediction and allocation.

4.3.1 Minimal Cut Sets (MCSs) and Minimal Path Sets (MPSs)

MCSs and MPSs are one of the principal tools of Design Risk Engineering.

4.3.1.1 Minimal Cut Sets (MCSs)

MCSs represent situations in which repair/improvement of any system component results in functioning/improvement of the system. Cut sets are the unique combinations of component failures that can cause system failure. Specifically, a cut set is said to be an MCS if, when any basic event is removed from the set, the remaining events collectively are no longer a cut set.

MCSs represent situations in which repair/improvement of any system component results in functioning/improvement of the system. MCSs coincide with circumstances under which failure/degradation of any system component causes system failure/degradation.

4.3.1.2 Minimal Path Sets (MPSs)

Path set is a set of components of a structure that by functioning ensures that the structure is functioning. MPS: A path set of a structure that cannot be reduced without losing status as a path set.

If all the components of a path set are functioning, the structure is functioning. The statement "cannot be reduced" means that if we remove one component from an MPS, the set is no longer a path set.

MCSs and MPSs allow us to compute a specific measure known as Fussell-Vesely Importance (FVI), which is used to evaluate importance of system components for system operation. The FVI has originally been developed for analysis of binary-state systems.

4.3.2 Wang Entropy's Applications to Design Risk Engineering

4.3.2.1 System Fault Diagnosis

For applications to system fault diagnosis, a fault tree diagnosis methodology which can locate the actual MCS in the system in a minimum number of inspections has been presented in Wang (1996) and Wang (2019a). The result reveals that, contrary to what is suggested by traditional diagnosis methodology based on probabilistic importance, inspection on a basic event whose FVI is nearest to 0.5 best distinguishes the MCSs.

4.3.2.2 System Reliability, Maintainability, and Testability Evaluation

For applications to system reliability, maintainability, and testability evaluation, a goal-tree-based methodology which can locate the actual MPS in the system in a minimum number of inspections has been developed. The result reveals that, contrary to what is suggested by traditional diagnosis methodology based on probabilistic importance, inspection on a basic event whose FVI is nearest to 0.5 best distinguishes the MPSs. This methodology has been extended to a dynamic goal tree approach and a dynamic fault tree approach.

4.3.2.3 Reliability Block Diagram (RBD) as a Tool for Design for Risk Engineering

An RBD is a diagrammatic method for showing how component reliability contributes to the success or failure of a redundant. RBD is also known as a dependence diagram. An RBD is drawn as a series of blocks connected in parallel or series configuration. RBDs are a way of representing a system, including its subsystems and components, as a series of blocks in such a way that equipment failure rates, operating philosophies, and maintenance strategies can be quantitatively assessed in terms of the impact they are expected to have on system performance.

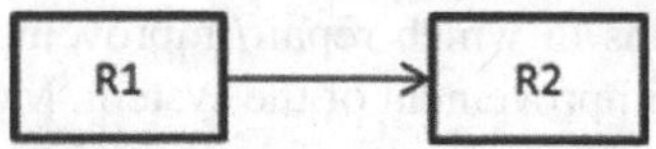

FIGURE 4.5 Simple series RBD.

The RBD is used to identify potential areas of poor reliability and where improvements can be made to lower the failure rates for the equipment. This method can be used in both the design and operational phases to identify poor reliability and provide targeted improvements.

The RBD shows the logical connections of components within a piece of equipment. It is not necessarily the schematic diagram of the equipment, but the functional components of the system. The equipment is made up of multiple components/systems in series, parallel, and a combination of the two. These components/systems and their configuration provide us with the inherent reliability of the equipment. The RBD analysis consists of reducing the system to simple series and parallel blocks.

4.3.2.3.1 Simple Series RBD

A simple series RBD is shown in Figure 4.5.

Minimal Path Set (MPS): {R1, R2}
Complexity Based on Wang Entropy: $\text{Log}_2(1) = 0$
Comments: for complexity: the number of inspections needed to identify the MPS leading to a system success is calculated to be zero based on Wang Entropy since only one MPS exists.

4.3.2.3.2 Simple Parallel RBD

A simple parallel RBD is shown in Figure 4.6.

Minimal Path Set: {R3}, {R4}
Complexity Based on Wang Entropy: $\text{Log}_2(2) = 1$
Comments: for complexity, the average number of inspections needed to identify the MPS leading to a system success is calculated to be one based on Wang Entropy since two MPSs exist.

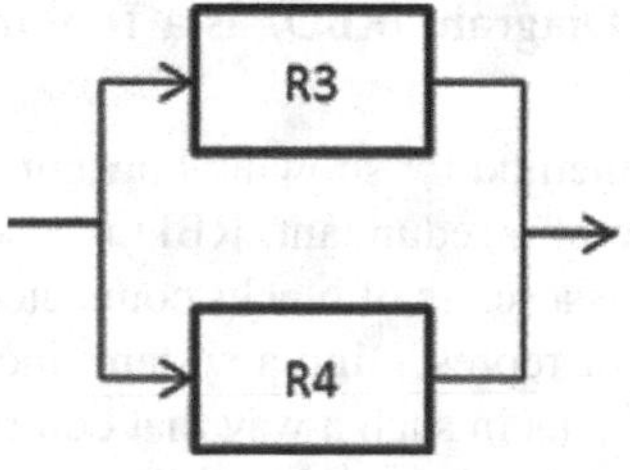

FIGURE 4.6 Simple parallel RBD.

Example 4.2

Evaluating complexity for 2-out-of-3 configuration if components are identical.

As shown in Figure 4.7, three hard drives in a computer system of Advanced Driver Assistance System (ADAS) are configured reliability-wise in parallel. At least two of them must function in order for the computer to work properly. Each hard drive is of the same size and speed; they are made by the same manufacturers and thus have the same reliability.

Since at least two hard drives must be functioning at all times, only one failure is allowed. This is a 2-out-of-3 configuration.

The following operational combinations are possible for system success:

- All three hard drives operate.
  - Path Sets: {HD #1, HD #2, HD #3}
- HD #1 fails, while HDs #2 and #3 continue to operate.
  - Path Sets: {HD #2, HD #3}
- HD #2 fails, while HDs #1 and #3 continue to operate.
  - Path Sets: {HD #1, HD #3}
- HD #3 fails, while HDs #1 and #2 continue to operate.
  - Path Sets: {HD #1, HD #2}

Thus, Minimal Path Sets: {HD #2, HD #3}, {HD #1, HD #3}, {HD #1, HD #2}

Complexity Based on Wang Entropy: $\text{Log}_2(3) = 1.585$

Comments: for complexity, the average number of inspections needed to identify the MPS leading to a system success is calculated to be 2 based on Wang Entropy since four MPSs exist.

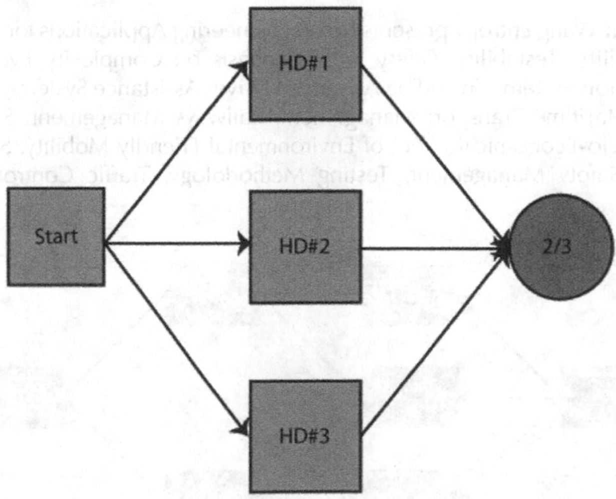

FIGURE 4.7 RBD showing three hard drives in a computer system of Advanced Driver Assistance System (ADAS).

Example 4.3

Application to Sustainable Transport Safety Management.

In many cases, it is not easy to recognize which components are in series and which are in parallel in a complex system. The network for Sustainable Transport Safety Management shown in Figure 4.8 is a good example of such a complex system. All the components A, B, C, D, E, F, and G have the same reliability.

The system in Figure 4.8 cannot be broken down into a group of series and parallel systems. This is primarily due to the fact that component C has two paths leading away from it, whereas B and D have only one.

- If C fails, Minimal Path Sets: {A, B, E, G}, {A, D, F, G}
- If C functions, Minimal Path Sets: {A, C, E, G}, {A, C, F, G}

Thus, combining the MPSs for C fails and C functions, MPSs for the system:

$$\{A,B,E,G\},\{A,D,F,G\},\{A,C,E,G\},\{A,C,F,G\}$$

Complexity Based on Wang Entropy: $\text{Log}_2(4) = 2$

Comments: for complexity, the average number of inspections needed to identify the MPS leading to a system success is calculated to be 2 based on Wang Entropy.

Optimal sequence of inspections: first inspect C;

1. then if C fails, inspect B, D, E, or E to distinguish Minimal Path Sets {A, B, E, G} and {A, D, F, G}.
2. if C functions, inspect E or F to distinguish Minimal Path Sets {A, C, E, G} and {A, C, F, G}.

Conclusion: Wang Entropy presents broad Engineering Applications for Reliability, Maintainability, Testability, Safety with emphasis on Complexity Evaluation of Transportation Systems, including Advanced Driver Assistance Systems, Intelligent Vehicles, Maritime Transport Management, Railways Management, Smart Card Reader, Socio-Economical Effect of Environmental Friendly Mobility, Sustainable Transport Safety Management, Testing Methodology, Traffic Control Devices,

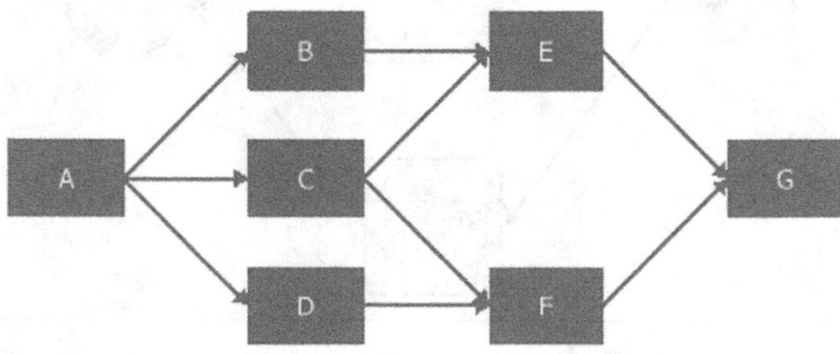

FIGURE 4.8 Network for Sustainable Transport Safety Management.

Traffic Psychology, Traffic Signal Priority System, Transport Management, Transport Safety, and Positive Train Control (PTC).

4.4 FMEA: FAILURE MODE AND EFFECT ANALYSIS

Failure is a big problem. We may touch it from different perspectives and get an increasingly complete picture. Because of the rapidly changing customer expectations and increasing regulations, industry's need for a disciplined approach to identify and prevent potential problems is more important now than ever before. Failure Mode and Effect Analysis (FMEA) is a powerful design assurance technique used to identify and minimize the effects of potential problems in a product or process design. In its most rigorous form, an FMEA is a summary of the engineer's thoughts (including the analysis of every conceivable item that could go wrong based on experience and past problems) as he/she designs a component or system. This systematic approach parallels and formalizes the mental discipline that an engineer normally goes through in a design process.

FMEA is a tool used during the formal design review stage by people who are responsible for the design, manufacture, management, and maintenance of the product. Their collective knowledge, experience, and ideas are harnessed to answer these fundamental questions:

- How might this product/process potentially fail to meet its intent?
- What might be the cause and effect of such failure?
- What controls do we have in place to detect the failure?
- What safety features might prevent the failure?

Consideration is then given to the potential failure modes and their implications to the design and manufacturing processes. This allows appropriate countermeasures to be developed so that high-risk components are designed to minimize the likelihood of that failure. FMEA is a technique for eliminating costly failures in the design/manufacture of a product by ensuring that critical issues are addressed before expensive commitments have been made. The product of an FMEA is a table of information that summarizes the analysis of all possible failure modes. In the following paragraph, we carefully look at the elements of this analysis.

4.4.1 Failure

Our first step in developing the FMEA technique is to discuss the meaning of failure. A failure simply means a component or system not meeting or not functioning to design intent. The parameters of the design intent might be stated in lifetime/cycles, linear dimensions/tolerances, load or deflection, coat thickness, etc. A failure attribute might also include static occurrences such as sharp edges, too glossy a finish, etc. Using this definition, a failure does not need to be readily detectable by a customer to still be considered to be a failure.

The idea of failure spreads across all of the various engineering fields. Engineering products are successful only to the extent that their designers properly anticipate how a product can fail to perform as intended. Virtually, every calculation that an engineer performs in the development of a ship and airplane, steam boiler or nuclear power reactor, is a failure calculation. From a broader perspective, the failure criterion is "it does not satisfy the customer's need".

Example 4.4

As shown in Figure 4.9, shortly after the upper floors of a high-rise hotel had been renovated to increase the hotel's room capacity, the guests complained that the elevators were too slow. The building manager assembled his assistants. His instructions to satisfy the perceived customer's need: *Find a way to speed up the elevators.* After calling the elevator company and an independent expert on elevators, it was determined that nothing could be done to speed up the elevators. Next, the manger's directions were *"Find a location and design a shaft to install another elevator"*. An architectural firm was hired to carry out this request. However, neither the shaft nor the new elevator was installed because shortly after the firm was hired, an alternative solution was uncovered. The *real need* was to find a way to reduce the impatient wait and one solution was to take the guests' minds off their wait rather than to install more elevators. The guests stopped complaining when mirrors were installed on each floor in front of the elevators.

4.4.2 Failure Mode

The FMEA technique starts by identifying the possible failure modes. A failure mode is the manner in which a component or system failure occurs; it is the manner in which the part or system does not meet design intent. The failure mode is the answer to the question, "How could the component or system fail"?

An aluminum beverage can, which all of us have used, may be described as a pressure vessel when it contains a carbonated beverage, especially under accident conditions when it has been dropped or shaken. As shown in Figure 4.10, the can

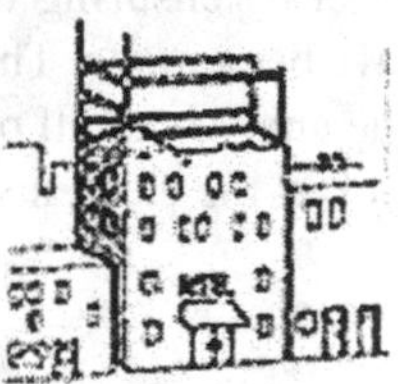

FIGURE 4.9 Failure – unable to satisfy customer's need.

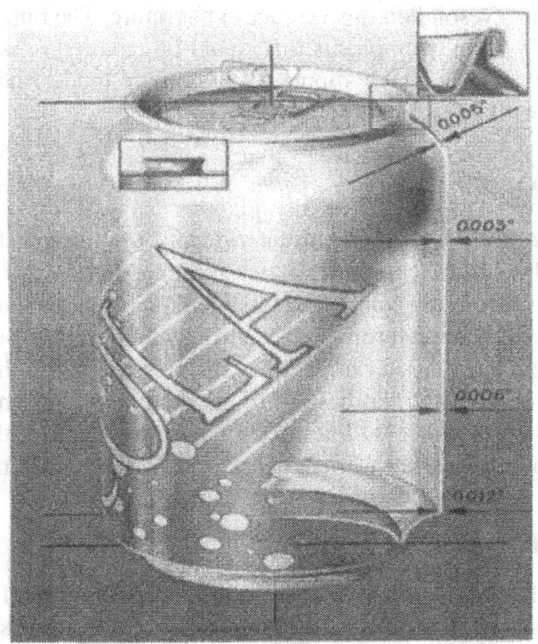

FIGURE 4.10 An aluminum can.

must be designed as carefully against accidental failure by explosion as does a steam boiler or nuclear reactor vessel, which has strict safety design codes. The successful design of a beverage can depends on avoiding the possibility that it will fail. The possible failure modes include:

1. The pressure of the compressed liquid splits the can open;
2. The compression of the top causes the stepped neck to be pushed down into the can;
3. The can's side wrinkles the way they do in an empty can;
4. The can's bottom pops out;
5. The can's bottom splits open;
6. The can's top arches to accommodate the pressure;
7. The rivet in the pop-top is ejected or split open, again similar to the pressure relief valve in a steam boiler;
8. The top cracks open where it is scored;
9. The can begins to leak around the rim where the top joins the sides.

The failure mode is a function of the component within a system. In general, each component in a system is analyzed to determine its possible failure modes (open, short, mechanical failure, etc.). Every part has numerous potential failure modes and theoretically, there is almost no limit to the depth one could go. Practically, there is a point of diminishing returns where the added cost exceeds the benefits derived. It is okay to combine similar failure modes if they have the same effect. They can always be separated for finer resolution if necessary in the future. The initial FMEA should include all of the system components that would be repaired or replaced during a maintenance action. Additional components and failure modes should be added as previously unrecognized failures occur.

4.4.3 Failure Mechanisms

Within the FMEA process, it is important to identify failure mechanisms. Failure mechanisms are the possible causes of failures. A failure mechanism analysis answers the question: "How could the component or system fail in this failure mode?"

For the aluminum can failure mode "the can's bottom pops out", the pressure within the thin-wall can plays a critical role. The pressure inside the "pressure vessel" tends to balloon out the flat bottom of an aluminum can, preventing it from sitting flat on a shelf or table. By finding the failure mechanism, the characteristic inwardly dished bottom was developed, to act like an arch against the pressure.

4.4.4 Failure Effect

The next step of the FMEA is to examine the failure effects. The effect of a component failure depends upon the function of the component in the system. Two valves may have the same part number but the effect of a failure will depend upon what the valve is controlling. Therefore, it is very important that each system component be assigned a unique symbol or designator that is completely independent from the part number. The system schematic is the key document used to determine the effect of a failure of a specific part, in a specific failure mode. The FMEA considers each part and determines the effect that each failure mode will have on the overall system as well as the environmental impact.

Before the invention of the stay-on tab, the removable tab-tops were used in beverage cans. By the early 1970s, it became evident that the removable pop-tops of beverage cans were creating an environmental crisis. The small, sharp, ringed tabs of aluminum were being disposed of by the billions and distributed on roadsides, in parks, and on beaches. Besides creating a litter problem, they were presenting dangerous hazards where recreation seekers went barefoot and where small children swallowed things not intended to be ingested. Especially on beaches, where the pop-tops were often too small to be caught in the standard beachcomber's rake, there was many a foot cut on an unseen tab lying just below the surface of the sand. The effect of these tabs on the environment should not have been unforeseen and resulted in the design failing. The failures of small pieces can accumulate to have a global impact.

4.4.5 Occurrence Ranking (O)

The previous steps of the FMEA were qualitative in nature and we discussed their relationship to the design process. The next steps are semi-quantitative and we will discuss the occurrence ranking, the severity ranking, and the detection ranking.

Occurrence is the likelihood that a specific cause/mechanism will occur. This is done by estimating the probability of occurrence on a scale "1" to "10". Any fail-safe controls intended to prevent the cause of failure and which are part of the current design should be considered in this estimate. When estimating the occurrence ranking, the following two probabilities should be considered:

1. The probability that the potential cause of the failure will occur. For this probability, all current fail-safe controls, which are in place to prevent the cause of failure from occurring on the part, must be assessed;
2. The probability that once the cause of the failure has occurred, it will result in the indicated potential failure mode. For this estimate, it must be assumed that the cause of failure and the failure mode are not detected before the product reaches the customer.

The engineer should mentally combine these two probabilities when estimating the occurrence ranking. The following occurrence ranking system should be used to ensure consistency.

- Remote – probability of occurrence. Unreasonable to expect failure to occur. (1)
- Low – failure rate. Related to similar designs having low failure rates (2 to 3)
- Moderate – failure rate. Related to similar designs that have occasional failures of the moderate type. (4 to 6)
- High – failure rate. Relates to failures in similar designs that have failed in the past. (7 to 9)
- Very high – failure rate. Almost certain that failure will occur in major way. (10)

To change the occurrence ranking for a particular design level, one of two actions must be taken:

- Change the design to reduce the probability that the cause of failure will result in the failure mode.
- Increase or improve the fail-safe "control systems"' which prevent the cause of failure from occurring.

Removing or controlling one or more of the causes/mechanisms of the failure mode through a design change is the only way a reduction in the occurrence ranking can be effected.

4.4.6 Severity Ranking (S)

Severity is an assessment of the seriousness of the effect of the potential failure mode to the next component, subsystem, system, or customer if it occurs. Severity applies to the effect only. Estimate the severity of the effects of failure to the customer on a "1" to "10" scale.

- Unreasonable – to expect the customer to notice the very minor failure. (1)
- Low Severity – ranking. Only slight customer annoyance. (2 to 3)

- Moderate – failure causing some customer dissatisfaction. Customer annoyed. (4 to 6)
- High – degree of failure resulting in the product not working and customer angry. (7 to 9)
- Very High – degree of failure. This rank indicates that the customer is at risk. Safety regulations are being infringed. (10)

A reduction in the severity ranking index can be effected only through a design change.

4.4.7 Detection Ranking (D)

Detection is an assessment of the ability of the proposed current design controls to identify a potential cause (design weakness) before the component, subsystem, or system is released for production.

- Unlikely – current design controls will/cannot detect a potential design weakness, or currently there are no design controls; (10)
- Very Low – current design controls probably will not detect a potential failure cause/mechanism (design weakness); (8 to 9)
- Low – current design controls not likely to detect a potential failure cause/mechanism (design weakness); (6 to 7)
- Moderate – current design controls may detect a potential failure cause/mechanism (design weakness); (4 to 5)
- High – current design controls have a good chance of detecting a potential failure cause/mechanism (design weakness); (2 to 3)
- Very High – current design controls will almost certainly detect a potential failure cause/mechanism (design weakness). (1)

As discussed above, the severity of a potential failure is represented by the variable S and is assigned a value between 1 and 10, where 10 is the most severe. The occurrence of the failure (Relative Failure Rate) is represented by the variable O and is assigned a value between 1 and 10, where 10 is the highest failure rate. The ability to detect a failure is represented by the variable D which is assigned a value between 1 and 10 with 10 being the most difficult to detect. The relative importance of a failure mode is represented by its Risk Product Number (RPN) which is calculated as:

$$RPN = S * O * D$$

The FMEA is an engineering technique used to define, identify, and eliminate known and/or potential problems from a system. The FMEA is an ongoing process that should start as a part of the first design review and continue throughout the life of the product. As shown in Table 4.1, the FMEA form provides a record of the various kinds of failures to which the system is subject. The FMEA database can be queried to generate special reports concerning system performance.

FMEA is a brainstorming, prioritization, and action item definition process performed by a multi-disciplined team from design, manufacturing, quality,

TABLE 4.1
Failure Mode and Effect Analysis of Aluminum Cans

Part	Failure Mode	Failure Mechanism	Effect	Frequency Ranking	Severity Ranking	Detection Ranking	RPN; Corrective Actions
Side	Break	Pressure of Compressed Liquid Splits Can's Sides Open	Leaking of Liquid	4	7	4	RPN = 112; Enough wall thickness to counter the internal pressure.
	Wrinkle	Insufficient Wall Stiffness	Bad Product Appearance	4	5	3	RPN = 60; Stiffen the thin-wall design.
Top	Stepped Neck Pushed Down into the Can	External Impact	Bad Product Appearance; Possible Leaking	3	6	6	RPN = 108; Impact protection during shipping.
	Arches	Pressure of Compressed Liquid Make the Can's Top Arch Out	Bad Product Appearance	3	5	3	RPN = 45; Stiffen the thin-wall design.
Bottom	Pops Out	Increased liquid pressure	Preventing can from sitting flat on a shelf or table	5	6	5	RPN = 150; Inwardly dished bottom acts like an arch dam.
	Splits Open	Pressure of Compressed Liquid Splits Can's Bottoms Open	Leaking of Liquid	3	7	4	RPN = 84; Enough wall thickness to counter the internal pressure.
Rivet in the Pop-Top	Ejected or Split Open	High Pressure When Dropped or Shaken	Leaking of Liquid	5	8	4	RPN = 160; Qualification under accident conditions

and field engineering. The goal of FMEA is to assure that the most critical design, manufacturing, and field failure modes of a product or process have been identified and addressed before they occur. FMEA is, therefore, an anticipatory process utilized to reduce overall costs and improve customer satisfaction. Corrective actions should be established for potential failure modes with significant RPNs. Sometimes administrative measures rather than design features should be implemented.

For years, FMEA has been an integral part of engineering design and has grown to be one of the most powerful and practical process control and reliability tools in manufacturing environments. Government agencies (i.e., Air Force and Navy) require that FMEAs be performed on their systems to ensure safety as well as reliability. And, for the most part, FMEA has become an indispensable tool for the aerospace and automobile industries.

Most notably, the automotive industry has adopted FMEAs in the design and manufacturing/assembly of automobiles. FMEA has been recently referenced in the Sematech FMEA Guide for Continuous improvement for the Semiconductor Equipment Industry (Sematech 92020963A) and in the European Union (EU) ISO 9004 quality standard as an effective tool for insuring quality product design and manufacture. The FDA recently issued a quality planning bulletin (FDA 90-4236) urging implementation of FMEA during pre-production phases.

FMEA and Failure Modes, Effects and Criticality Analysis (FMECA) are inductive techniques for analyzing a system to assess the risk of it failing during operational use. The process consists of analyzing the kinds of failure that are possible (to assess their likelihood), and determining the effects that each kind of failure would have if it were to occur during operational use (to assess their severity). The combination of likelihood and severity constitutes criticality, which indicates its importance in risk analysis. The kinds of failure, or failure modes, depend on the design and the implementation technology. When used in conjunction with goal tree analysis and other risk engineering techniques, FMEA is one of the most powerful tools available for identifying reliability, safety, compliance, and product non-conformities in the design stages rather than during the production process. FMEA is an engineering team effort where the team is dedicated to creating a robust, compliant, reliable product the first time out of the gate.

4.5 REDUNDANCY AND FAULT TOLERANCE

Engineered systems are developed to satisfy a set of requirements that meet customer needs. A requirement that is important in many engineering systems is that they be highly dependable. In the case of equipment used in space missions, satellite communications, complex military systems, etc., failure would have far-reaching consequences. For those engineering systems, redundancy is a means of achieving dependability. In general, redundancy means more than one way of accomplishing a required function. This means that part or the whole of particular vulnerable equipment is replicated in order to reduce the probability of loss of system function.

Redundancy can take many forms: partial, active, standby, etc. Partial redundancy can be illustrated by spokes in a bicycle wheel; even if some of them are broken, the wheel will not fail (see Figure 4.11). A lift suspension may consist of more than one rope so that in case one fails, the lift can still operate. A further example is that of a four-engine airplane, capable of taking off on three engines. Shared-load is a typical form of partial redundancy.

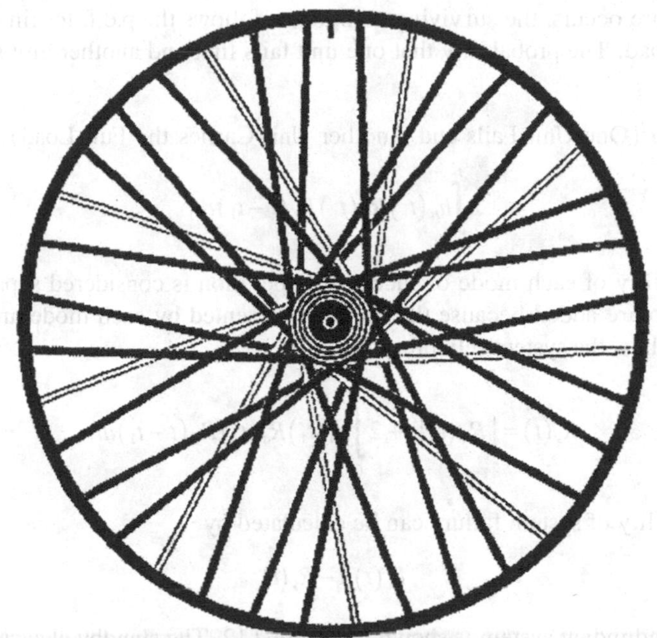

FIGURE 4.11 Bicycle spokes and partial redundancy.

In a shared load parallel configuration, the subsystems equally share the load initially. When a subsystem fails, the remaining subsystems must sustain an increased load. Therefore when more subsystems have failed, the failure rates of the surviving components increase. A simple example is two flashlight batteries placed in parallel to provide a fixed voltage. Assume the circuit is designed so that if either fails, the other will supply adequate voltage. Nevertheless, the current through the remaining battery will be higher, and this will cause greater heating in the internal resistance. The net result is that the remaining battery will operate at a higher temperature and thus tend to deteriorate faster.

Let's assume that the corresponding reliability functions during full-load and half-load operations are $R_f(t)$ and $R_h(t)$, respectively. The analysis simplifies the situation to only two subsystems, where

$f_h(t)$ = probability distribution function (p.d.f.) of time to failure distribution under full load;
$f_f(t)$ = p.d.f. of time to failure distribution under half-load;
$h_h(t)$ = hazard rate for unit under half-load.

Before a failure occurs, both units follow the p.d.f. for time to failure under half-load. The probability of both units working successfully is

$$P(\text{both units working under half load}) = \left[R_h(t)\right]^2$$

After a failure occurs, the surviving component follows the p.d.f. for time to failure under full load. The probability that one unit fails first and another unit switches to full load is

$$P(\text{One Unit Fails and Another Unit Carries the Full Load})$$

$$2\int_0^t h_h(t_1)R_h(t_1)R_f(t-t_1)dt_1$$

The probability of each mode of successful operation is considered separately and then the two are added because the events represented by each mode are mutually exclusive. Thus, the system reliability becomes

$$R_s(t)=\left[R_h(t)\right]^2+2\int_0^t h_h(t_1)R_h(t_1)R_f(t-t_1)dt_1$$

The probability of system failure can be calculated by

$$P_s(t)\,1-R_s(t)$$

A standby redundant system is shown in Figure 4.12. The standby electrical generator, system B, is used in a hospital to assure electrical power supply in the event of main generator A's failure. In the case of the failure of A, generator B can be switched into the system either manually or automatically. Here, the standby generator is only brought into operation when the normal operating generator fails. Suppose the probability of generator A failure is P_A and the probability of the generator B failure during standby is P_B; the probability of switch failure is P_S. The overall system fails if:

- Both generators A and B fail; the probability of this event combination is $P_A P_B$;
- Generator A fails and switch fails; the probability of this event combination is $P_A \bullet P_S$.

The probability of overall system failure is

$$P_{system}=P_A P_B+P_A \bullet P_S$$

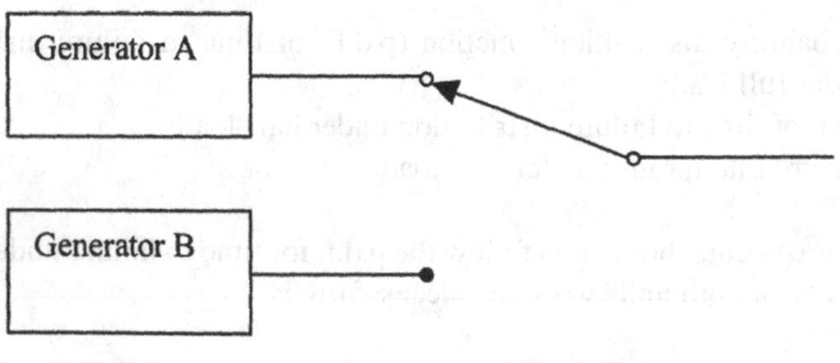

Generator A

Generator B

FIGURE 4.12 Standby redundancy.

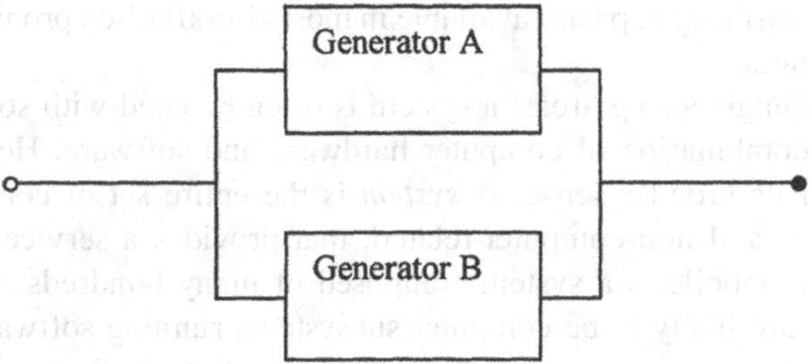

FIGURE 4.13 Parallel redundancy.

Parallel redundancy is another form of redundancy frequently employed for improving dependability. As shown in Figure 4.13, *A* and *B* are identical devices, both capable of performing their function independently. In the event of failure of either of the elements *A* or *B*, the overall system function will not be lost.

Here, only one individual generator is necessary to meet the functional requirements placed on the overall system. The overall system will only fail, if both generators *A* and *B* fail. Thus, the probability that the overall system fails is the probability that generator *A* fails and generator *B* fails. Suppose the probability of generator failure during normal operation is *P*. Assuming that the failure probability of individual generators is independent, the probability of the overall system failure can be calculated as:

$$P_{system} = P^2$$

Generally for a parallel redundant system consisting of *n* identical and independent components, the probability of overall system failure can be calculated by

$$P_{system} = P^n$$

This ignores dependency of failure of the redundant elements. In estimating the reliability of systems with redundant elements, it is critical to estimate the impact of dependencies that are observed to have a significant impact in all redundant systems.

Many other redundancy configurations are possible; for example, we might consider a parallel system of three or more elements where the overall function is unaffected by the loss of one element, but the loss of two elements would result in degradation of function but not complete loss.

It is obvious that the use of redundancy techniques increases the size and cost of an engineering system due to the duplication of system elements. This limits its application in many engineering areas where redundancy is highly desirable, but in practice is frequently prevented by space/weight considerations. For example, the transmitter in an airborne radar system is usually the area of greatest reliability concern because of the particularly severe environment associated with high power (hence high temperature) and high voltage combined with vibration. The limited space and load-carrying capacity available in most aircraft often prohibit the replication of transmitters.

In the many engineering areas, a system is often equated with software, or perhaps with the combination of computer hardware and software. Here, we use the term system in its broader sense. A *system* is the entire set of components, both computer-related and non-computer-related, that provides a service to a user. For instance, an automobile is a system composed of many hundreds of components, some of which are likely to be computer subsystems running software. Fault tolerance is a means of achieving dependability for hardware/software-integrated systems. There are three levels at which fault tolerance can be applied.

Traditionally, fault tolerance has been used to compensate for faults in computing resources (hardware). By managing extra hardware resources, the computer subsystem increases its ability to continue operation. *Hardware fault tolerance* measures include redundant communications, replicated processors, additional memory, and redundant power/energy supplies. Hardware fault tolerance was particularly important in the early days of computing, when the time between machine failures was measured in minutes.

A second level of fault tolerance recognizes that a fault-tolerant hardware platform does not, in itself, guarantee high availability to the system user. It is still important to structure the computer software to compensate for faults such as changes in program or data structures due to transients or design errors. This is *software fault tolerance*. Mechanisms such as checkpoint/restart, recovery blocks, and multiple-version programs are often used at this level.

At a third level, the computer subsystem may provide functions that compensate for failures in other system facilities that are not computer-based. This is *system fault tolerance*. For example, software can detect and compensate for failures in sensors. Measures at this level are usually application-specific. It is important that fault tolerance measures at all levels be compatible; hence, the focus is on system-level issues in this document.

Hazards to systems are a fact of life. So are faults. Yet, we want our systems to be dependable. A system is dependable when it is trustworthy enough that reliance can be placed on the service that it delivers. For a system to be dependable, it must be available (e.g., ready for use when we need it), reliable (e.g., able to provide continuity of service while we are using it), safe (e.g., does not have a catastrophic consequence on the environment), and secure (e.g., able to preserve confidentiality).

Although these system attributes can be considered in isolation, in fact they are interdependent. For instance, a system that is not reliable is also not available (at least when it is not operating correctly). A secure system that doesn't allow an authorized access is also not available.

Achieving the goal of dependability requires effort at all phases of a system's development. Steps must be taken at design time, implementation time, and execution time, as well as during maintenance and enhancement. At design time, we can increase the dependability of a system through *fault avoidance* techniques. At implementation time, we can increase the dependability of the system through *fault removal* techniques. At execution time, *fault tolerance* and *fault evasion* techniques are required.

4.5.1 Fault Avoidance

Fault avoidance uses various tools and techniques to design the system in such a manner that the introduction of faults is minimized. A fault avoided is one that does not have to be dealt with at a later time. Techniques used include design methodologies, verification and validation methodologies, modeling, and code inspections and walk-through.

4.5.2 Fault Removal

Fault removal uses verification and testing techniques to locate faults enabling the necessary changes to be made to the system. The range of techniques used for fault removal includes unit testing, integration testing, regression testing, and back-to-back testing. It is generally much more expensive to remove a fault than to avoid a fault.

4.5.3 Fault Tolerance and Fault Evasion

In spite of the best efforts to avoid or remove them, there are bound to be faults in any operational system. A system built with fault tolerance capabilities will manage to continue operating, perhaps at a degraded level, in the presence of these faults. For example, a fault-tolerant power system comprises multiple power converters configured to maintain the integrity of the output power bus in the event of any one single-point failure within the power conversion system. Fault-tolerant power systems can be designed in a variety of configurations, but all share the following common traits:

- Sufficient capacity to sustain power bus operation in the event of any single-point power system fault (redundancy);
- Ability to isolate and localize a failure to a single replaceable module (fault isolation and detection);
- Design that permits extraction of the faulty module and insertion of a replacement without interruption of the power bus (on-line replacement, or "hot-swap").

The process of fault-tolerant design provides a means to achieve a balanced project risk where the cost of failure protection is commensurate with the program resources and the mission criticality of the equipment. By providing compensation for potential hardware/software failures, a fault-tolerant design approach may minimize engineering risk without recourse to non-optimized redundancy or over-design.

As shown in Figure 4.14, the fault-tolerant design presumes that the potential causes of failure are identifiable. This identification is performed by hardware/software FMEA. Similarly, fault tree analysis (FTA) and event tree analysis (ETA) identify safety issues and potential faults in mechanical and electromechanical devices. Engineers utilize the results of FMEA, FTA, and ETA to establish fault-tolerant equipment design priorities. Fault-tolerant design is an iterative process; its current validity relies on the current iteration of the FMEA and FTA/ETA.

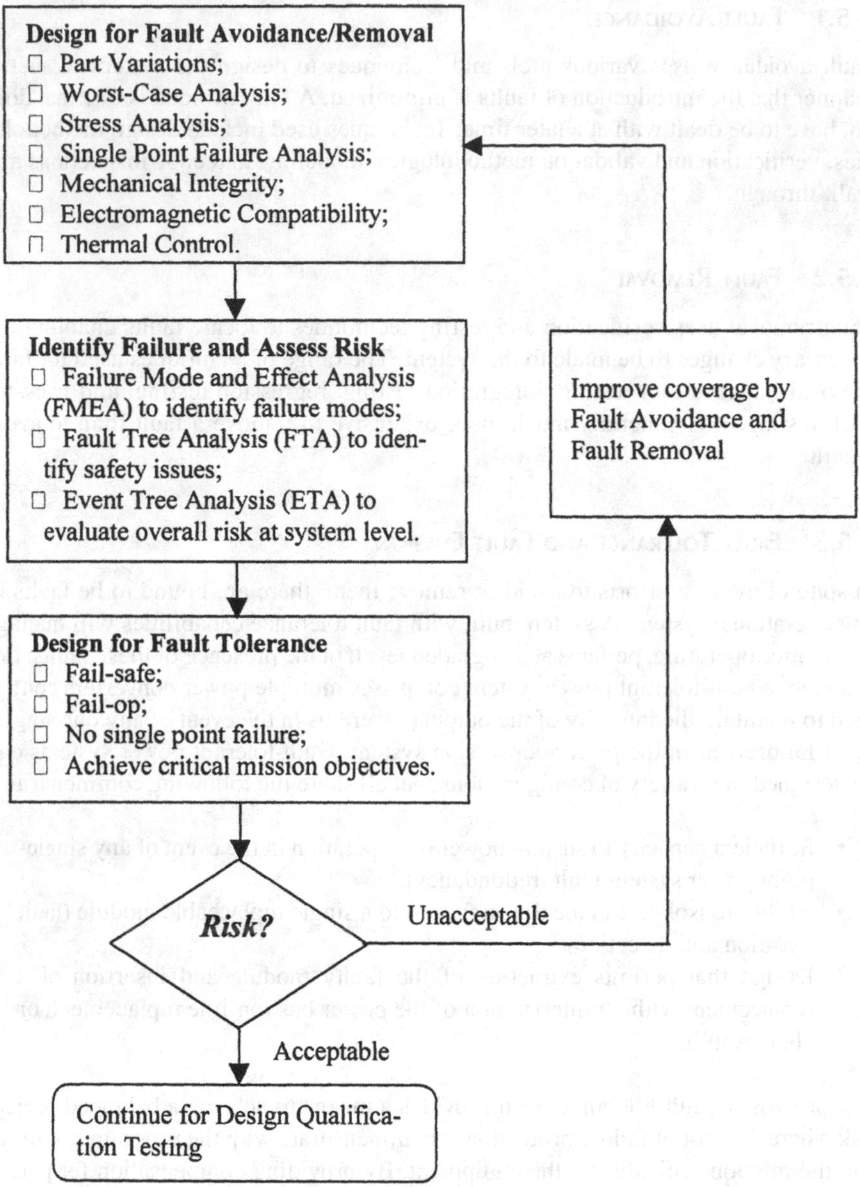

Design for Fault Avoidance/Removal
- ☐ Part Variations;
- ☐ Worst-Case Analysis;
- ☐ Stress Analysis;
- ☐ Single Point Failure Analysis;
- ☐ Mechanical Integrity;
- ☐ Electromagnetic Compatibility;
- ☐ Thermal Control.

Identify Failure and Assess Risk
- ☐ Failure Mode and Effect Analysis (FMEA) to identify failure modes;
- ☐ Fault Tree Analysis (FTA) to identify safety issues;
- ☐ Event Tree Analysis (ETA) to evaluate overall risk at system level.

Design for Fault Tolerance
- ☐ Fail-safe;
- ☐ Fail-op;
- ☐ No single point failure;
- ☐ Achieve critical mission objectives.

Risk?

Unacceptable

Improve coverage by Fault Avoidance and Fault Removal

Acceptable

Continue for Design Qualification Testing

FIGURE 4.14 Process flow diagram for fault-tolerant design.

The fault-tolerant design approach is based upon the expectation that failures will occur. However, their effects will be automatically counteracted by incorporating either redundancy or other types of compensation. A fault-tolerant design approach differs from a pure design redundancy approach in that provisions are made for planned degraded modes of operation where acceptable.

Example 4.5

Fail-op

Design the system so that, when it sustains a specified number of faults, it still provides a subset of its specified behavior. The high gain antenna of a spacecraft is usually not redundant because of its size. A fault-tolerant design would favor the use of a backup medium gain antenna operating at reduced data rates as a degraded but acceptable operating mode. Similarly, a partially failed power source within a solar panel array or a failure of one of three radioisotope thermoelectric generators (RTGs) would be accommodated: an appropriate failure detection circuit and a software fault protection algorithm could be provided to shed low-priority electrical loads or instruments, while maintaining the most important mission capabilities.

Example 4.6

Fail-safe

Design the system so that, when it sustains a specified number of faults, it fails in a safe mode. For instance, railway signaling systems are designed to fail so that all trains stop. Similarly, the reactor safety protection system is designed as fail-safe: given loss of power supply, the control-rods with neutron-absorbing materials will be inserted into the reactor core automatically to stop the nuclear reaction inside the reactor; emergency cooling will also be provided to remove the residual heat within the reactor core.

System failures occur when faults propagate to the outer boundary of the system. The goal of fault tolerance is to intercept the propagation of faults so that failure does not occur, usually by substituting redundant functions for functions affected by a particular fault. Occasionally, a fault may affect enough redundant functions that it is not possible to reliably select a non-faulty result and the system will sustain a *common-mode failure.*

A common-mode failure results from a single fault (or fault set). Common-mode failures are caused by phenomena that create dependencies between two or more redundant components that cause them to fail simultaneously. Such failures have the potential for negating much of the benefit gained with redundant configurations. Common-mode failures may be caused by common electric connections, shared stresses such as dust or vibration, common maintenance problems, or a host of other factors. For example, computer systems are vulnerable to common-mode resource failures if they rely on a single source of power, cooling, or I/O. A more insidious source of common-mode failures is a design fault that causes redundant copies of the same software process to fail under identical conditions.

The most effective fault-tolerant approach for combating common-mode design errors is design diversity – the implementation of more than one variant of the function to be performed. For computer-based applications, it is generally accepted that it is more effective to vary a design at higher levels of abstraction (i.e., by varying the algorithm or physical principles used to obtain a result) than to vary implementation details of a design (i.e., by using different programming languages or low-level coding techniques). Since diverse designs must implement a common system specification, the possibility for dependencies always arises in the process of refining the specification to reflect difficulties uncovered in the implementation process. Truly diverse designs would eliminate dependencies on common design teams, design philosophies, software tools and languages, and even test philosophies. Many approaches attempt to achieve the necessary independence through randomness, by creating separate design teams that remain physically separate throughout the design, implementation, and test process. Recently, some projects have attempted to create diversity by enforcing differing design rules for the multiple teams.

Example 4.7

The on-board shuttle software runs on two pairs of primary computers, with one pair being in control as long as the simultaneous computations on both agree with each other, with control passing to the other pair in the case of a mismatch. All four primary computers run identical programs. To prevent catastrophic failures in which both pairs fail to perform (for example, if the software were wrong), the shuttle has a fifth computer *that is programmed with different code by different programmers from a different company,* but using the same specifications and the same compiler (HAL/S). Cut-over to the backup computer must be done manually by the astronauts.

A fault-tolerant design can provide dramatic improvement in system reliability and leads to a substantial reduction in technical risk as a consequence of fewer disabling system failures. The essential ingredients to achieving a fault-tolerant design are the performance of a thorough FMEA and FTA/ETA, the detailed communication of these identified failure modes and effects among engineering teams. Presentation of fault-tolerant design options to management requires a skilled engineering team with intimate knowledge of system operation and clear understanding of the weight, volume, schedule, and cost constraints.

4.6 RISK ENGINEERING AND INTEGRATING INFORMATION SCIENCE

In previous chapters/sections, a fault tree diagnosis methodology which can locate the actual MCS in the system in a minimum number of inspections is presented. An entropy function is defined to estimate the information uncertainty at a stage of diagnosis and is chosen as an objective function to be minimized. Inspection which can provide maximal information should be chosen because it can minimize the information uncertainty and will, on average, lead to the discovery of the actual MCS in a minimum number of subsequent inspections. The result reveals that, contrary to what is suggested by traditional diagnosis methodology based on probabilistic importance, inspection on a basic event whose FVI is nearest to 0.5 best distinguishes the MCSs.

FTA is widely used in reliability/risk assessment. It is also used as a tool for diagnosis. Wang (1996) and Wang (2019a) presented an optimal sequence of diagnosis and repair for a two-state repairable system, in which the order of inspections (on the states of components) is based upon the ranking of the ordinary and contrary measures. If at a certain stage of diagnosis a component is found to be in failure, maintenance action is taken to repair the component. After that, if the system is still in failure, continue the diagnosis and repair process. Such an approach is not well formulated because it confuses diagnosis and repair. Diagnosis is an information-gathering process to find the actual MCS occurring in the system, and thus recover the system at minimum cost (time, money, etc.). An optimal repair strategy can be devised only after a well-conducted diagnosis. In this section, we concentrated on the diagnosis problem. We consider diagnosis as a process which:

1. identifies all the MCSs C; (i = 1, 2, …, n) with their importances I;
2. proposes information-gathering queries to locate the actual MCS at the lowest cost (e.g., minimum number of inspections).

Here, we concentrate on the Fussell-Vesely basic event importance for discussion because it is widely accepted as a well-defined one. As discussed in Chapter 1, the FVI of the basic event B_i is defined as the probability that at least one MCS containing B_i has failed, given that the top event has failed.

4.6.1 Diagnosis: A Process of Minimizing Wang Entropy

Wang Entropy, quantities of the form $W_H = -\sum I_i log_2 I_i$ play a central role in the diagnosis methodology as a measure of uncertainty. Diagnosis is an information gathering process which discriminates among the original MCSs. Such a process minimizes the information uncertainty associated with a diagnosis problem, i.e., minimizes the Wang Entropy. We are certain of the diagnosis results if and only if we know from the available information that all the I but one are zero, this one having the value unity, i.e., if and only if $W_H = -\sum I_i log_2 I_i = 0$. It is ideal to minimize this entropy to zero. In practice, a threshold value, W_T, can be defined. Diagnosis is terminated if $W_H \leq W_T$.

In this section, the concept "information", which reflects the expected decrease in the information uncertainty after an inspection, is introduced to evaluate the effect of an inspection. Criteria for choosing a next inspection, which will provide maximal information and thus will, on average, lead to the discovery of the actual MCS in a minimum number of subsequent inspections, are developed. A diagnosis algorithm is devised as follows:

1. Identify all the MCSs C, with their corresponding importance I;
2. Select the basic event to be inspected according to the criteria whose FVI is nearest to 0.5.
3. Update cut set importance I_i according to the inspection outcome (see Chapter 1). After inspecting the basic event *Bi*, whose FVI is nearest to 0.5, the actual MCS can be located with the minimum expected number of additional.
4. Calculate Wang Entropy W_H by the updated cut set importances. If $W_H \leq W_T$, end diagnosis and output diagnosis result. Otherwise, go to step (2) and continue diagnosis.

4.6.2 Maximal Information: The Criteria for Selecting Inspection

Inspections provide information to the diagnostician which helps to differentiate among the MCSs. At this stage of diagnosis, the inspection that provides maximal information should be taken next because such an inspection is expected to minimize the entropy function and thus minimize the information uncertainty involved. This inspection will, on average, lead to the discovery of the actual MCS in a minimum number of subsequent inspections.

4.6.3 Conclusion

This section reveals that Wang Entropy, an entropy function of MCS importances, as a measure of information uncertainty associated with a diagnosis problem, is the minimum expected number of inspections needed to identify the actual MCS. The concept of information provided by an inspection, which indicates the expected decrease in the minimum expected number of inspections to identify the actual MCS after this inspection has been performed, is introduced. From the research work, we find the following results:

Cut set importance is very useful for fault tree diagnosis. Based on this cut set importance, an entropy function can be devised to express the information uncertainty at a stage of diagnosis. A diagnosis can be formulated as a process to minimize this entropy function. Inspection that can provide maximal information should be chosen because it will, on average, lead to the discovery of the actual MCS in a minimum number of subsequent inspections. Fussell-Vesely basic event importance is a useful concept for diagnosis, but its use is not as straightforward as in risk analysis and risk reduction. To identify the actual MCS, the basic event with medium FVI should be selected to inspect.

BIBLIOGRAPHY

Anon. (1989), *Risk Management Concepts and Guidance*, Defense Systems Management College, Fort Belvoir, VA.

Batson, R. G. (1987), "Critical Path Acceleration and Simulation in Aircraft Technology Planning," *IEEE Transactions on Engineering Management*, Vol. EM-34, No. 4, pp. 244–251.

Batson, R. G. and Love, R. M. (1988), "Risk Assessment Approach to Transport Aircraft Technology Assessment," *AIAA Journal of Aircraft*, Vol. 25, No. 2, pp. 99–105.

Bell, T. E., ed. (1989), "Special Report: Managing Risk in Large Complex Systems," *IEEE Spectrum*, June, pp. 21–52.

Beroggi, G. E. G. and Wallace, W. A. (1994), "Operational Risk Management: A New Paradigm for Decision Making," *IEEE Transactions on Systems, Man, and Cybernetics*, Vol. 24, No. 10, pp. 1450–1457.

Black, R. and Wilder, J. (1979), "Fitting a Beta Distribution from Moments," Memorandum, Grumman, PDM-OP-79-115.

Book, S. A. and Young, P. H. (1992), "Applying Results of Technical-Risk Assessment to Generate a Statistical Distribution of Total System O θ 8 ζ," *Presented at the AIAA 1992 Aerospace Design Conference*, Irvine, CA, February 3–6.

Chapman, C. B. (1979), "Large Engineering Project Risk Analysis," *IEEE Transactions on Engineering Management*, Vol. EM-26, No. 3, pp. 78–86.

Chiu, L. and Gear, T. E. (1979), "An Application and Case History of a Dynamic R&D Portfolio Selection Model," *IEEE Transactions on Engineering Management*, Vol. EM-26, No. 1, pp. 2–7.

Cullingford, M. C. (1984), "International Status of Application of Probabilistic Risk Assessment," *Risk & Benefits of Energy Systems: Proceedings of an International Symposium*, Vienna, Austria, IAEA-SM-273/54, pp. 475–478.

Dean, E. B. (1993), "Correlation, Cost Risk, and Geometry," *Proceedings of the Fifteenth Annual Conference of the International Society of Parametric Analysts*, San Francisco, CA, June 1–4.

Dienemann, P. F. (1966), *Estimating Cost Uncertainty Using Monte Carlo Techniques*, The Rand Corporation, Santa Monica, CA, January, RM-4854-PR.

Dodson, E. N. (1993), *Analytic Techniques for Risk Analysis of High-Technology Programs*, General Research Corporation, Santa Barbara, CA, RM-2590.

Fairbairn, R. (1990), "A Method for Simulating Partial Dependence in Obtaining Cost Probability Distributions," *Journal of Parametrics*, Vol. X, No. 3, pp. 17–44.

Garvey, P. R. and Taub, A. E. (1992), "A Joint Probability Model for Cost and Schedule Uncertainties," *Presented at the 26th Annual Department of Defense Cost Analysis Symposium*, September.

Greer, W. S. Jr. and Liao, S. S. (1986), "An Analysis of Risk and Return in the Defense Market: Its Impact on Weapon System Competition," *Management Science*, Vol. 32, No. 10, pp. 1259–1273.

Hazelrigg, G. A. Jr. and Huband, F. L. (1985), "RADSIM - A Methodology for Large-Scale R&D Program Assessment," *IEEE Transactions on Engineering Management*, Vol. EM-32, No. 3, pp. 106–115.

Henley, E. J. and Kumamoto, H. (1992), *Probabilistic Risk Assessment*, IEEE Press, Piscataway, NJ.

Hertz, D. B. (1979), "Risk Analysis in Capital Investment," Harvard Business Review, Vol. 9, pp. 169–181.

Honour, E. C. (1994), "Risk Management by Cost Impact," *Proceedings of the Fourth Annual International Symposium of the National Council of Systems Engineering*, Vol. 1, San Jose, CA, August 10–12, pp. 23–28.

Hutzler, W. P., Nelson, J. R., Pei, R. Y., and Francisco, C. M. (1985), "Nonnuclear Air-to-Surface Ordnance for the Future: An Approach to Propulsion Technology Risk Assessment," *Technological Forecasting and Social Change*, Vol. 27, pp. 197–227.

Keeney, R. L. and von Winterfeldt, D. (1991), "Eliciting Probabilities from Experts in Complex Technical Problems," *IEEE Transactions on Engineering Management*, Vol. 38, No. 3, pp. 191–201.

Markowitz, H. M. (1959), *Portfolio Selection: Efficient Diversification of Investment*, Second Edition, John Wiley & Sons, New York, NY, reprinted in 1991 by Basil Blackwell, Cambridge, MA.

McKim, R. A. (1993), "Neural Networks and the Identification and Estimation of Risk," *Transactions of the 37th Annual Meeting of the American Association of Cost Engineers*, Dearborn, MI, July 11–14, pp. P.5.1–P.5.10.

Ock, J. H. (1996), "Activity Duration Quantification under Uncertainty: Fuzzy Set Theory Application," *Cost Engineering*, Vol. 38, No. 1, pp. 26–30.

Quirk, J., Olson, M., Habib-Agahi, H., and Fox, G. (1989), "Uncertainty and Leontief Systems: An Application to the Selection of Space Station System Designs," *Management Science*, Vol. 35, No. 5, pp. 585–596.

Roush, M. L., Modarres, M, and Hunt, R. N. (1985), "Application of Goal Trees to Evaluation of the Impact of Information upon Plant Availability," Proceedings of the International ANS/ENS Topical Meeting on Probabilistic Safety Methods and Applications, San Francisco.

Roush, M. L. and Wang, J. X. (1995), "Time-Dependent Logic for Goal-Oriented Dynamic System Analysis," *Proceedings of 1995 Annual Reliability and Maintainability Symposium*, Washington, DC, January 16–19.

Savvides, S. (1994), "Risk Analysis in Investment Appraisal," *Project Appraisal*, Vol. 9, No. 1, pp. 3–18.

Sholtis, J. A. Jr. (1993), "Promise Assessment: A Corollary to Risk Assessment for Characterizing Benefits," *Tenth Symposium on Space Nuclear Power and Propulsion*, American Institute of Physics Conference Proceedings 271, Part 1, Albuquerque, NM, pp. 423–427.

Shumskas, A. F. (1992), "Software Risk Mitigation," in Schulmeyer, G. G. and J. I. McManus (eds.), *Total Quality Management for Software*, Van Nostrand Reinhold, New York, NY.

Skjong, R. and Lerim, J. (1988), "Economic Risk of Offshore Field Development," in *Transactions of the American Association of Cost Engineers*, American Association of Cost Engineers, New York, NY, pp. J.3.1–J.3.9.

Timson, F. S. (1968), *Measurement of Technical Performance in Weapon System Development Programs: A Subjective Probability Approach*, The Rand Corporation, Santa Monica, CA, Memorandum RM-5207-ARPA.

Wang, J. X. (1996), "Complexity as a Measure of the Difficulty of System Diagnosis," *International Journal of General Systems*, Vol. 24, No. 3, pp. 257–269.

Wang, J. X. (2017), *Industrial Design Engineering: Inventive Problem Solving*, CRC Press, Boca Raton, FL.

Wang, J. X. (2019a), "Complexity as a Measure of the Difficulty of System Diagnosis in Next Generation Aircraft Health Monitoring System," SAE Technical Paper 2019-01-1357, doi:10.4271/2019-01-1357.

Wang, J. X. (2019b), "A Dynamic Fault Tree Approach for Time-Dependent Logical Modeling of Autonomous Flight Systems," SAE Technical Paper 2019-01-1358, doi:10.4271/2019-01-1358.

Williams, T. (1995), "A Classified Bibliography of Recent Research Relating to Project Risk Management," *European Journal of Operations Research*, Vol. 85, pp. 18–38.

5 Risk Acceptability
Uncertainty in Perspective

5.1 UNCERTAINTY: WHY BRIDGES FALL DOWN

Uncertainty denotes a degree of ignorance about engineering risk, i.e., ignorance about the following three questions:

What can go wrong within an engineering system?
How likely is the failure to happen?
What will be caused by the failure as a consequence?

The 2,800-feet main span of the Tacoma Narrows Bridge was the third-longest suspension bridge when it was opened in July 1940. In designing the slender bridge, engineers utilized a method that was quite accurate for determining lateral deflections of the truss and cable stresses in the truss due to lateral forces. The bridge was well designed and built to resist safely all static conditions according to the state of the art. The constant sideways push of the wind was taken into account in the standard way of that time.

However, as shown in Figure 5.1, the aerodynamic forces of the turbulent wind and the potential to excite resonance of the structure were ignored by the engineers. The new field of aerodynamics was being applied to the development of airplanes in the 1930s, but was seen to be irrelevant to designing and analyzing large static structures like bridges. Nobody recognized that this aerodynamic force would dominate the behavior of the light, only two-lane wide and very shallow, deck of the Tacoma Narrows Bridge.

The ignored failure mode during engineering design was doomed to lead to later surprising consequences. Even before the Tacoma Narrows Bridge was completed, engineers were surprised by its large movement. On November 7, 1940, the Tacoma Narrows Bridge executed its fatal oscillations (see Figure 5.2). The motion was so violent that the massive steel towers were permanently bent out of shape and the bridge had to be dismantled before a replacement bridge could be built.

Engineering products are as good as the engineer's ability to foresee potential failures. Engineering uncertainty in the face of the unforeseen – or the unforeseeable – was ultimately what brought the Tacoma Narrows Bridge down. Ignorance about a hidden failure mode was also the underlying cause in 1967 when the seemingly sturdy, 40-year-old Silver Bridge over the Ohio River at Point Pleasant, Ohio, collapsed without warning, plunging 46 motorists to their deaths.

The collapse of the Silver Bridge proved to be such an enigma to investigators that the study continued for four years. Using state-of-the-art technology, the investigators finally determined the reason: metal fatigue had caused one I-beam to crack and

DOI: 10.1201/9781003371014-5

FIGURE 5.1 Bridge experienced an ignored failure mode.

FIGURE 5.2 Collapse due to the ignored failure mode.

TABLE 5.1
Types of Uncertainties

Types of Uncertainties	Descriptions
Inherent Uncertainty	Due to variability inherent in the material or the environment. Example: Cycles-to-failure in fatigue have large variability as observed in fatigue tests.
Statistical Uncertainty	Resulting from incompleteness of statistical data (small sample sizes).
Modeling Uncertainty	Resulting from assumptions made in analysis of stresses and strengths, i.e., the use of simplified models. Example: A large number of assumptions are made in the process of estimating stress at a notch in a component, given environmental conditions.
Human Error	Errors in calculation; Chose wrong known data; Inadequate design review; Failed to calculate critical conditions; Poor quality fabrication; Wrong materials used; Poor judgment and abuse by operators.

fail, bringing the rest of the bridge down with it. The fatal crack was so thin that it had not been detected in the standing structure. The sensitivity of bridge failure to this single element was a lesson learned from this detailed analysis.

Engineers, constructors, and maintainers may create an unforeseen failure mode themselves, as we could observe from the collapse of the West Gate Bridge at Melbourne, Australia, in 1970. Trying to connect the main lengthwise splice of the bridge, engineers started removing bolts from the main transverse splice at mid-span to correct for misalignment. Before they had taken out enough bolts to correct it, they removed so many that the bridge collapsed.

In a close examination of a typical design problem, one observes an extremely complicated process because of the many uncertainties that exist in an engineering analysis. The four major types of uncertainties, inherent uncertainty, statistical uncertainty, modeling uncertainty, and human error, are summarized in Table 5.1. The structural failures discussed above illustrate the need for engineers to understand risk and uncertainty.

5.2 RISK MITIGATION: HOW BUILDINGS STAND UP

Many people marvel at engineers' abilities to design buildings that stretch toward the sky, but few realize that without engineers' inventions for minimizing engineering risk, previous generations would never have dreamed of building these great landmarks.

Risk Mitigation: Risk mitigation uses measures that are introduced into system design and operation to reduce the probability or consequence of undesirable events even when there are system failures.

Elisha Otis was born in Halifax, Vermont. As a mechanical engineer, he was inspired to design what came to be called the "safety elevator" when he was asked

Failure Modes:

1. **Lift cylinder fracture;**
2. **Lifting rope break;**

Risk Mitigation

Engaged teeth on the guide-rail locked the falling elevator car.

FIGURE 5.3 Risk mitigation against loss of lift.

to move equipment into the warehouse of his employer, a New York bed factory. Elevators in use at that time were operated by hydraulic power. Breaking of the lifting rope or a fracture of the hydraulic lift cylinder often caused catastrophic consequences on these early elevators. Otis' employer needed an elevator that could carry people and equipment safely to the upper floors of its new building.

Later, at the Crystal Palace Exposition in New York in 1853, Otis demonstrated his engineering solution. A large crowd watched breathlessly from the floor far below as Otis ascended in his new elevator. Stopping at a dizzying height, Otis told his assistant to cut the elevator's cord.

The crowd let out a gasp of relief when the elevator platform did not come crashing to the floor. As shown in Figure 5.3, the key to Otis' invention was a toothed guide-rail located on each side of the elevator shaft that caught the elevator car. If the cable failed, the teeth would engage, locking the car in place.

Risk mitigation in this example was accomplished by an engineered feature that provided a safety backup. In some other situations, risk mitigation may be most effectively accomplished by administrative controls that will reduce the frequency of exposure to potential harm.

Risk Prevention: Engineering risk should be eliminated and "designed out" of the product or system, if possible. By removing the failure mechanisms, the failure modes can be practically eliminated from the engineering systems.

After Otis' successful introduction of the first safety-brake-equipped elevator in the 1850s and the introduction of steel-frame construction, buildings began to grow upward. In 1910, the Metropolitan Life building broke all records for height until that time: it was 50 stories high.

With the construction of the 102-story Empire State Building in 1931, skyscrapers began to sprout in many cities worldwide. Engineers began to challenge what had been accepted as the "traditional" method of designing and constructing skyscrapers. Innovations in skyscraper design such as lighter materials, increased window area, and cantilevered supports resulted in taller, lighter, and slimmer buildings.

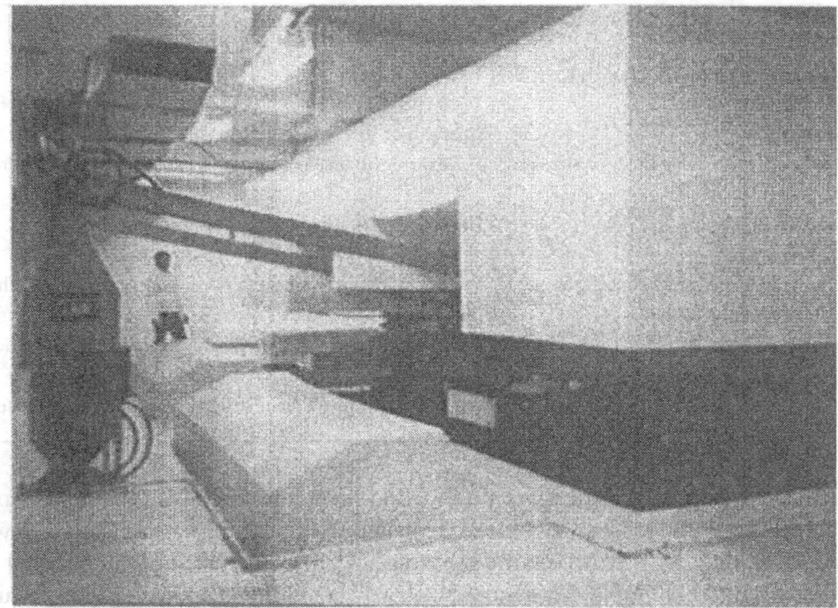

FIGURE 5.4 Tuned-mass damper: Eliminating oscillations.

The Citicorp tower in New York was constructed in 1977 using a diagonal bracing design, which translated into great weight savings. The tower was unusually light for its size. However, this meant that it would have a tendency to sway in the wind, so a *tuned-mass damper* was installed at the top of the building (Figure 5.4).

In this case, it was not necessary to stiffen the structure to reduce the oscillations as was done for the new Tacoma Narrows Bridge nor to increase its strength to overcome the fatigue of large oscillations. A means to avoid swaying of the building was engineered into the structure (see Figure 5.5).

FIGURE 5.5 Building swaying in the wind.

TABLE 5.2
Load and Strength Uncertainties

Load or Stress	Strength or Resistance
The assumptions used in the load/stress analysis contain error.	The exact resistance properties of the material are unknown.
The magnitudes of the peak loads are not exactly known.	The size effects are not accurately known.
Discontinuities and stress concentrations can only be approximated.	The effects of machining and processing operations on the resistance are not known.
Overload by operators.	There is uncertainty of the effect of the assembly operations on the resistance of the system.
	The effect of time on the resistance is difficult to evaluate.

Robust Design: Buildings are designed to resist the forces to which they may be subjected. However, the resistance or material strength is not perfectly known. In addition, the forces to which the building structure may be subjected cannot be predicted precisely. Table 5.2 summarizes load and strength uncertainties, which greatly influence engineering risk. Robust design minimizes the sensitivity to material strength and applied forces.

A tube derives its resistance to bending from the fact that the main supporting materials are concentrated around the periphery rather than at the center of the structure. The neutral axis is approximately at the center of the structure; therefore, it has very low stresses. Buildings like the twin towers of the World Trade Center or those of the John Hancock Center in Chicago are often called *tube buildings* since their outer walls act as the walls of a hollow tube. They are the most efficient structures designed so far against the wind. The John Hancock Center, shown in Figure 5.4, has a weight of only 29.7 pounds per square foot of floor, while the Empire State Building has a total weight of 42.2 pounds per square foot of floor.

Wind loads as high as 45 pounds per square foot (120 mph) were considered in the structural design of the World Trade Center towers. As have become standard with tall buildings, model tests were conducted in a wind channel in order to confirm theoretical and computer calculations. There were also considerations of how the wind blowing between the towers would affect their behavior.

Engineers have acquired a great amount of knowledge about risk minimization for high-rise buildings. That is why skyscrapers continue to stand.

5.3 FROM SAFETY FACTOR TO SAFETY INDEX

Whenever an item is subjected to varying stresses or loads, it will continue to function properly as long as the item has adequate strength. Failure will occur if the load exceeds the strength. Load and strength are considered here in the broadest sense. "Load" may refer to mechanical stress, an electrical voltage, or thermal stress such as temperature. "Strength" may refer to any resisting physical property, such as hardness, strength of material, fatigue strength, melting point, or adhesion.

To ensure that a system will not fail, it is a standard practice to design all parts to have greater strength than the load that is anticipated. It is common to characterize this over-design in terms of a safety factor which is calculated by dividing the load required to cause failure by the maximum load expected to act on an engineering product.

$$\text{Safety Factor} = \text{SF} = \frac{\text{Load Causing Failure}}{\text{Max. Anticipated Load}}$$

$$= \frac{S}{L} \tag{5.1a}$$

The critical failure mode in a column is likely to be buckling. In the case of the pyramids where their shape makes buckling virtually impossible, compression is the critical failure mode upon which the safety factor should be based. A column of marble, concrete, or limestone could be as high as 12,000 feet before collapsing in compression under its own weight. Stronger stone, like granite, could reach 18,000 feet. But the tallest stone buildings ever erected, the limestone pyramids of Egypt, are less than 500 feet. Prudent ancient engineers incorporated a factor of safety of at least 24 in their pyramid designs.

Due to the uncertainties shown in Table 5.2, neither load nor strength are fixed quantities. For example, consider a rope with a rating of 6,000 pounds being used in a hoist which is rated to lift 1,000 pounds at a time. For this rope, the factor of safety is 6. However, even if the hoist is rated at 1,000 pounds, operators may overload it by 50%. Thus, a 1,500-pound load could be on the hoist at any given time. Furthermore, even though the operator is instructed to start and stop the hoist smoothly, it may be operated in a jerky manner occasionally. Due to the effects of inertia, the load on the rope might be further increased from 1,500 pounds to 3,000 pounds.

In addition, the rope strength may be less than the specified 6,000 pounds due to several reasons. An inferior rope may have been installed unwittingly. The rope may have become frayed and weakened due to use or abuse. If the net effects were that the rope could barely support 3,000 pounds, the jerky lifting of an overloaded hoist bucket could break the rope. The hoist could fail due to uncertainties and variabilities even with a factor of safety of 6.

It now becomes evident that we cannot tell whether a given safety factor is adequate unless we also know how much uncertainty or variability we have. Probabilistic distributions are used to describe uncertainties. Both load and strength are variables, distributed about their mean values, μ_L and μ_s. Standard deviations, σ_L and σ_s, are directly related to the uncertainties, the dispersions of load and strength, and provide a well-defined measure of these variations.

Let's use G to denote the separation of strength and load, which we will assume are both normally distributed. The item will not fail as long as G is positive, i.e., the strength exceeds the load.

$$G = S - L \tag{5.1b}$$

Then, the mean value and standard deviation of G are as follows:

$$\mu_G = \mu_S - \mu_L \tag{5.2}$$

And

$$\sigma_G = \left(\sigma_S^2 + \sigma_L^2\right)^{1/2} \tag{5.3}$$

σ_G denotes the combined standard deviation of the load and strength distributions. This measure is very important since it is the combined effect of uncertainties in strength and uncertainties in load.

We now chose to use the safety index shown below as the separation of the mean values of strength and load in terms of the amount of uncertainty.

$$SI = \mu_G / \sigma_G \tag{5.4}$$

Example 5.1

Load and strength are normally distributed (units in thousand-pounds):

Load ~ Normal(20,10);
Strength ~ Normal(40,10).

Calculate the safety factor and safety index of the structural design.

The safety factor is calculated by dividing the mechanical strength (40 thousand-pounds) by the expected load (20 thousand-pounds). The safety factor is 2.

By using Equations (5.2) and (5.3):

$$\mu_G = \mu_S - \mu_L = 20$$

And

$$\sigma_G = \left(\sigma_s^2 + \sigma_L^2\right)^{1/2} = 14.1$$

From Equation (5.4), the safety index is calculated as:

$$SI = \mu_G / \sigma_G = 1.4$$

As shown in Figure 5.6, any item in the left-hand tail of the strength distribution (items with smaller than average strength) has some chance of failure because they "overlap" the right-hand tail of the load distribution (items with larger than average load). The probability of failure is significant since there is a large overlap between load and strength.

Example 5.2

The implementation of an overload protection system has now reduced the standard deviation of the load distribution. In addition, the improved material control process has reduced the standard deviation of the strength distribution. Again, the load and strength are normally distributed (units in thousand-pounds):

Load ~ Normal(20,4);
Strength ~ Normal(40,4).

Calculate the safety factor and safety index of the structural design.

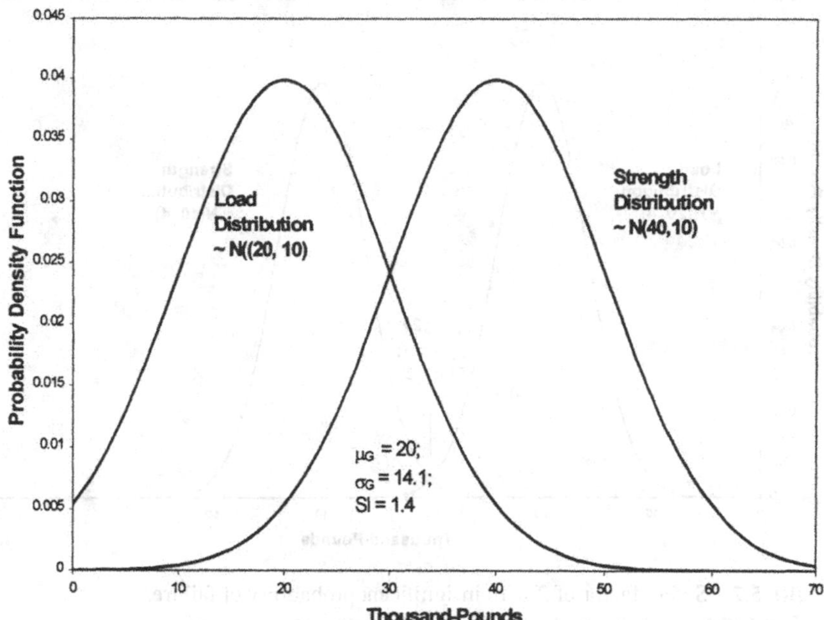

FIGURE 5.6 Safety factor of 2 with significant probability of failure.

Obviously, the safety factor is still 2 since the mechanical strength and the expected load remain 40 thousand-pounds and 20 thousand-pounds, respectively.

The mean value of G, μ_G stays at 20 according to Equation (5.2). Based on Equation (5.3), σ_G is calculated as follows:

$$\sigma_G = (\sigma_S^2 + \sigma_L^2)^{1/2} = 5.7$$

From Equation (5.4), the safety index is increased to

$$SI = \mu_G / \sigma_G = 3.5$$

This indicates a considerable increase in safety because now the amount that S_{avg} exceeds L_{avg} (mean of G) is 3.5 times the size of the standard deviation of G.

As shown in Figure 5.7, the probability of failure is significantly reduced from that in Example 5.1 due to the greatly reduced overlapping area. Only when an item at the extreme weak end of the strength distribution is subjected to a load at the extreme high end of the load distribution will failure now occur.

We will find in Section 5.4 that this reduction of variability reduces the probability of failure from 0.081 to 0.00023. Thus, the reduction of variability to 40% of its previous values has reduced the load-strength-overlap (probability of failure) by 350.

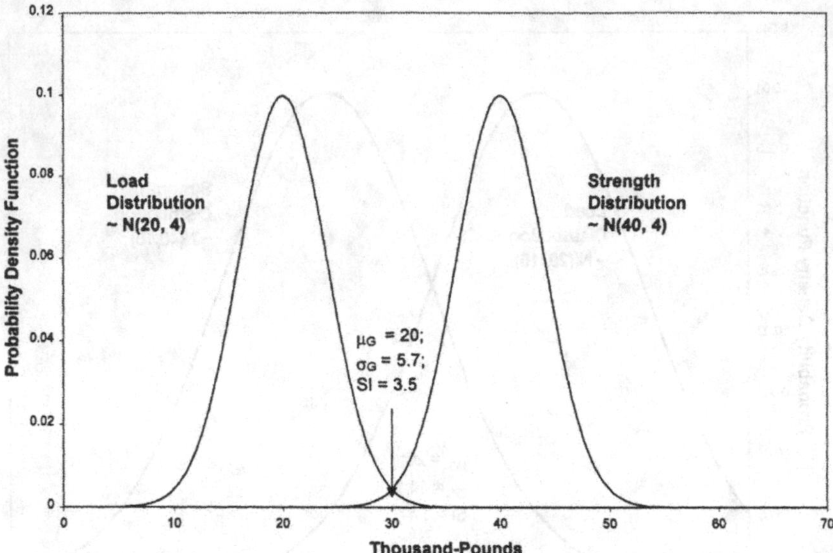

FIGURE 5.7 Safety factor of 2 with insignificant probability of failure.

The following average safety indices have been estimated for marine structural systems:

• Merchant-ship deck collapse	5.3
• Naval frigate deck collapse	2.2
• UK offshore structures	3.7
• US offshore structures	2.3
• Semi-submersibles	4.4
• North Sea tension leg platform	5.3
• UK steel bridges	4.8

The safety index is now being used as the basis for risk acceptance since it incorporates uncertainties of load and strength into the traditional safety factor concept. A minimum allowable safety index, β_0, can be specified as the design criteria for a component or a system. For example, the following target safety indices are recommended by *A. S.* Veritas Research, a Norwegian agency that certifies large-scale structures worldwide (Table 5.3).

Failure Type:

1. Ductile failure with reserve strength capacity resulting from strain hardening;
2. Ductile failure with no reserve capacity;
3. Brittle fracture and instability.

TABLE 5.3
Target Safety Indices for Mechanical Structures

Failure Consequence	Type 1 Failure	Type 2 Failure	Type 3 Failure
Not Serious	3.09	3.71	4.26
Serious	3.71	4.26	4.75
Very Serious	4.26	4.75	5.20

Failure Consequences:

Not Serious: A failure implying small possibility for personal injuries; the possibility for pollution is small; the economic consequence is small.
Serious: A failure implying possibilities for personal injuries/fatalities or pollution or significant economic consequences.
Very Serious: A failure implying large possibilities for several personal injuries/fatalities or significant pollution or very large economic consequences.

The safety index is the basis of the "Load and Resistance Factor Design" method specified by the American Institute of Steel Construction (AISC), the American Society of Civil Engineers (ASCE), and the American Association of Highway and Transportation Officials (AASHTO).

5.4 CONVERTING SAFETY INDEX INTO PROBABILITY OF FAILURE

We have indicated that the safety index is a useful quantity because it is related to the degree of safety for a particular design. We now want to find a quantitative connection between the safety index and the probability of failure. The probability of failure is the probability that the load exceeds strength. If $G = S - L$, we have:

$$P_f = P(G < 0) \tag{5.5}$$

If we consider normally distributed load and strength, then

$$\mu_G = \mu_S - \mu_L$$

And

$$\sigma_G = \sqrt{\left(\sigma_S^2 + \sigma_L^2\right)}$$

So,

$$P_f = \phi(-\mu_G/\sigma_G) \tag{5.6}$$

Therefore, given the safety index, the probability of failure can be determined by finding the value of the standard cumulative normal variate from the normal

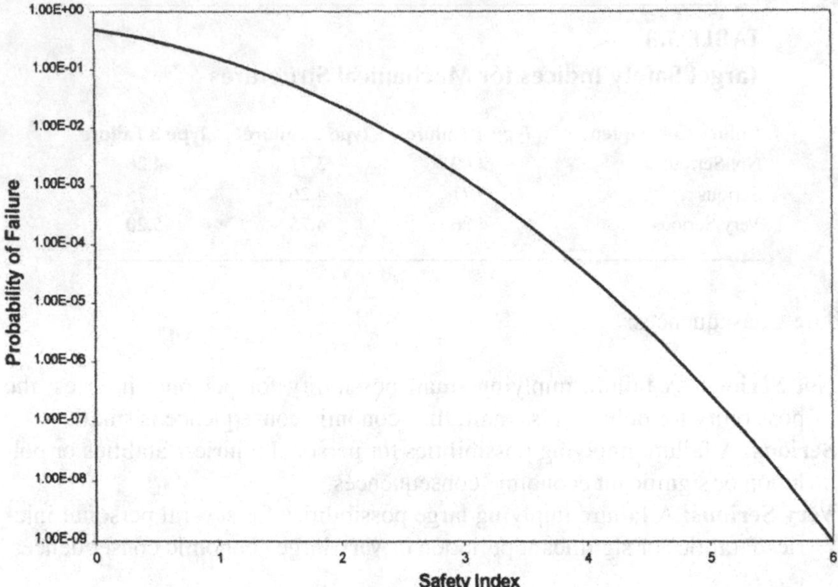

FIGURE 5.8 Probability of failure vs. safety index.

distribution tables. As shown in Figure 5.8, the probability of failure decreases as the safety index increases.

For example, the probability of failure is 0.081 when the safety index is 1.4; the probability of failure decreases to 0.00023 when safety index increases to 3.5. The probability of failure is a quantitative manifestation of Murphy's Law – it tells us how likely the failure is to occur. A maximum allowable probability of failure, P_0, can be specified as the basis for risk acceptance.

Example 5.3

For its internal combustion engine, a company issues an unconditional warranty on the crankshaft for 300,000 miles. Based on the maximum amount of money that the company can afford for warranties, the target probability of failure is specified as 0.003 for a crankshaft. The six-cylinder engine has two fatigue-failure sites (at the crank pin) per cylinder; thus, there are 12 sites for fatigue, each having equal risk and designed for a safety index of 3.5. Can the company accept the risk associated with this design considering their target probability of crankshaft-failure?

Since the safety index for each fatigue-site is 3.5, the probability of failure is

$$P_f = \phi(-3.5) = 2.3E-4$$

There are 12 sites for fatigue, with each having equal risk. The probability of crankshaft-failure is then:

$$P_{crankshaft} = 12 \times 2.3E-4 = 0.00276$$

Since the probability of failure is smaller than the target, the risk is acceptable with respect to fatigue design.

Back in 1960, the International Civil Aviation Organization (ICAO) recommended 10^{-5} as a maximum value for the probability of catastrophic failure in the service life of an aircraft, which initiated the use of quantitative safety goals for aircraft systems. More recently, the following probability-based risk criteria have been used by the French aeronautical industry:

- 10^{-9}/hour (extremely improbable) *Catastrophic Accidents.* The faults lead to the loss of the aircraft and/or numerous fatalities.
- 10^{-7}/hour (extremely rare) *Critical Accidents.* The faults may result in a dangerous reduction of the safety margins, in an excessive workload or physical stress, which may prevent the crew from performing all its tasks or cause serious injuries.
- 10^{-5}/hour (rare) *Significant Accidents.* The faults can lead to a perceptible reduction of safety margins, to a marked deterioration of the airworthiness or an appreciable increase in the crew's workload.

Similarly, the U.S. Federal Aviation Administration (FAA) suggested a goal of 10^{-9} for probability of catastrophic events occurring during a flight (including take-off and landing).

5.5 QUANTITATIVE SAFETY GOALS: PROBABILITY VS. CONSEQUENCE

For offshore oil rigs, major accidents have included the blow-out (uncontrolled gushing of hydro-carbon) on the Bravo oil production platform, helicopters falling on offshore platforms, and more recently the destruction of the Alexander L. Kielland oil rig. These accidents have made operators, authorities, and the public aware of the safety problems of operating such structures. All parts of society struggle to establish what is an acceptable risk as viewed from their particular perspective.

The acceptance of engineering risk involves both probability and consequence. There are two basic aspects of risk control:

- **Prevention:** lower the probability of the undesired event;
- **Mitigation:** make the consequences less unpleasant when the undesired event does happen.

For offshore oil rigs, the major risk-mitigation functions include:

- *The main structure*: In case of an accident, the main structure must continue supporting its load for a given period.
- *Areas of refuge*: These structures must remain intact several hours after the occurrence of an accident until a totally safe evacuation can be carried out.
- *Emergency exits*: At least one emergency exit must remain undamaged for at least one hour.

Quantitative safety goals have been developed to provide goals against which to measure the capability of risk prevention and mitigation functions. Setting safety goals requires establishing what levels of risk are acceptable to the public. Studies of public acceptance of risk have found that the public reaction to an accident is related to the following factors:

- *Basic risk level*: Inescapable minimum level of risk by death accepted by an individual member of society, e.g., $P' \approx 10^{-4}$/yr for industrial countries.
- *Social criterion factor*: If an activity is voluntary, a person may be willing to increase his/her exposure to danger by a factor K_s, e.g., typical value, $K_s = 5$.
- *Aversion factor*: Public reaction to an accident is presumably proportional to the number of people involved (n_r). This "aversion" is accounted for by dividing the acceptable risk by n_r; e.g., $n_r = 200$ on an offshore platform.

As shown in Figure 5.9, the maximum acceptable probability of failure for mechanical structures can be expressed as follows:

$$P_0 = (P'K_s)/n_r \tag{5.7}$$

For example, the safety goal for an offshore platform can be determined to be

$$P_0 = \left(1\times10^{-4}\times5\right)/200 = 2.5\times10^{-6}/\text{yr}$$

If the platform is designed for $N = 40$ years, the probability of failure during its design life should be less than

$$P_n = N \times P_0 = 10^{-4}$$

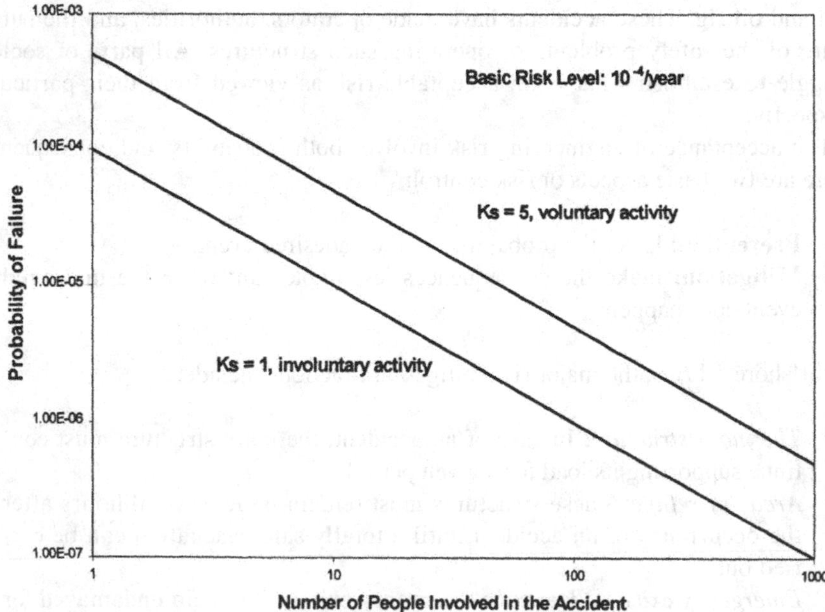

FIGURE 5.9 Risk acceptance for mechanical structures.

Designers of offshore oil platforms should use the following procedures to implement safety goals:

- Identify all the probable accidents and their initiating events; the probable accidents may include blow-outs, fires, objects falling, collision with ships, helicopter crashes, earthquakes, and extreme weather conditions.
- Assess occurrence probability for each probable accident identified above.
- Develop accident sequences following each initiating event; assess the probability of each accident sequence.
- Evaluate the consequence of each accident sequence, including its impact on the risk-mitigation functions. Accident sequences jeopardizing the risk-mitigation functions are called *residual accident events.*
- If the total probability of the residual events exceeds 10^{-4} per year, the design should be revised to meet the safety goal.

This approach of quantitative safety goals has been used by the Norwegian Petroleum Directorate for determining the design-basis of offshore rigs. The Canadian Standards Association (CSA) is considering the following design codes for offshore installations in Canadian waters:

- **10^{-5}/year:** *Safety Class 1.* Failure results in a great loss of life or a high potential for environmental damage.
- **10^{-3}/year:** *Safety Class* 2. Failure would result in small risk to life and a low potential for environmental damage.

For nuclear power plants, the U.S. Nuclear Regulatory Commission (NRC) issued both qualitative and quantitative safety objectives in 1986. The qualitative safety objectives are as follows:

- The public in the vicinity of the facilities should be adequately protected against potential consequences so that no individual might incur a significant additional risk to his life or health due to nuclear power plant operation.
- The collective risk to life and health arising from nuclear power plant operation should be comparable to or below the risks from power production by alternate feasible techniques and should not contribute to increase other collective risks significantly.

The quantitative safety objectives are as follows:

- The individual immediate death probability in the neighborhood of a plant site due to nuclear power plant accidents should not exceed one-thousandth of the cumulative immediate death probabilities from other accidents the American population is generally exposed to.
- The individual cancer-induced death probability (in the population living near a nuclear plant site) which may result from the plant operation should not exceed one-thousandth of the cumulative death probability imputable to cancers induced by all other causes.

- When considering actions that would reduce the exposure of some persons to radiation, the benefits for society from the additional reduction in mortality should be compared with the financial cost on the basis of 1,000 dollars per man-rem exposure reduction.
- The probability of a nuclear power plant accident resulting in an extensive core melt should remain below 10^{-4} per reactor year.

5.6 DIAGNOSABILITY: ADDITIONAL ELEMENT OF RISK

5.6.1 Diagnosability

Diagnosability is the study of a model's capacity to identify faults and distinguish between various internal fault types. Diagnosability is the property of a partially observable system with a given set of possible faults that can be detected with certainty with a finite observation.

5.6.2 Diagnosability Plan: Ensure the Diagnosability of a Partially Controllable System

A diagnosability plan specifies a sequence of applicable actions that leads the system from an initial belief state (a set of potentially current states) to a diagnosable belief state, in which the system is then left to run freely. The objective of a diagnosability plan is to ensure the diagnosability of a partially controllable system.

5.6.3 Design for Diagnosability

The centralization of process controls in contemporary workplaces is increasing the adoption of diagnosability in cutting-edge technology. Modern machinery must now perform diagnostics, which involves having enough sensory data and computing capability to not only identify the defect but also to pinpoint its fundamental cause. A final homogenous package must be created by integrating various third-party components, ranging in diagnosability from a straightforward digital or analog signal to a collection of alarms reflecting one or more failure modes.

5.6.3.1 Preventative Maintenance

Critical assets need time-based maintenance to restore their condition to the nominal value; however, this maintenance has a price that may be avoided with reliable diagnosability. By taking proper care of the machine through time-based and usage-based maintenance, wear-out failures may be avoided or reduced. Condition-based maintenance intervenes when early signs are identified and stops a decrease in functioning. As long as the designers can provide early failure detection, it can replace time-based maintenance. Both situations fall under the category of preventative maintenance because no component ever ceases to function, making it possible to limit the failure mode and minimize damage to other components.

5.6.3.2 Modularization

Modularization in diagnosability also suggests that FW checks are independent of the machine. The typical method for doing that is to implement the diagnostics checks as low in the stack as possible and then report the results to higher levels for failure isolation. This system cannot be followed by machine-wide failure modes; instead, higher levels must periodically or upon request receive the sensing data. In these situations, robust communication techniques that can maintain transmissions in safety or emergency situations and recover from them are extremely valuable.

5.6.3.3 Poka-yoke Design

Any failure should not have disastrous effects on the user or the machine, but engineers must always bear this in mind. Furthermore, interactions with people may result in unanticipated user behavior, which can result in unreported failures. In the case of a sufficiently big sample, precautions must be taken to prevent major issues. As previously mentioned, preventive measures like good robustness and diagnosability play a significant role in that regard, but the idea of poka-yoke design goes a little further.

To make its designs foolproof, the Japanese manufacturer Toyota was the first to implement this approach to the industry. Poka-yoke basically aims to design the process so that errors may be found and fixed right away, reducing flaws at the source.

Applications include the detection of a passenger's seatbelt being fastened in a car and connectors that can only be inserted one way, like a USB Type-A. Defects in the industry can be avoided by using the proper poka-yokes, deterring errors with warnings, or eliminating physical incompatibilities.

The majority of potential failures can be lessened in severity or perhaps avoided by planning with this paradigm in mind.

5.6.3.4 Failure Mode Effects Analysis (FMEA)

An FMEA should be performed at the early stages of the design process to visualize weak points or potential failures and develop an action plan to tackle them. It should be utilized to locate any missing sensory data that would improve the coverage of diagnosability. Moreover, it is suggested to reassess the risks related to each potential failure mode if the design is considerably altered. To ensure that corrective actions are implemented to minimize the most serious concerns discovered, it is crucial to monitor the action plan created during the FMEA.

5.6.3.5 Component Test

The low levels of the system check pyramid are finished after the Hardware (HW) identification and the power and status checks. The high levels begin with component test checks, which aggregate all failure modes that only involve one component and need a single sensor to identify them.

5.6.3.6 Status Signal's Redundancy

The only way to obtain sensor diagnostics for some feedbacks and status signals is to duplicate them and compare the redundant data sent to the controller. Switches merely close or open a circuit and their output can be represented by a single bit of information regarding their diagnosability. However, in important applications where a failure supervision is prohibited, designers might mandate them.

5.6.3.7 Components and Sensors with Built-in Communication Protocols

Numerous sensors need signal processing and amplification using certain ADCs, MCU pins, etc. However, operating components typically call for specific control systems. The majority of manufacturers also provide fully integrated alternatives, which include all of this functionality in the component's package and communicate with the rest of the system through a communication protocol (such I2C).

These sensors and components will take less time to construct because they are packaged, but they can also constrain the design because of their larger size, fewer features, communication needs, or diagnostic checks (the diagnostics integrated by the manufacturer may not cover the same range of failure modes the designers would have chosen). Additionally, they are frequently pricey and inappropriate for applications requiring high volume.

Low-volume products gain the most from the widespread adoption of these parts or sensors in the industry. Manufacturers may permit compatibility with various brands or models, allowing for flexibility and facilitating simple upgrades or replacements for faulty or outdated parts.

These components can use error codes or fault signals for diagnosability; however, the system engineer will need to comprehend them.

5.6.3.8 Summary

The centralization of process controls in contemporary workplaces is increasing the adoption of diagnosability in cutting-edge technology. Modern machinery must now perform diagnostics, which involves having enough sensory data and computing capability to not only identify the defect but also to pinpoint its fundamental cause. A final homogenous package must be created by integrating various third-party components, ranging in diagnosability from a straightforward digital or analog signal to a collection of alarms reflecting one or more failure modes.

Current methodologies and processes to enable good troubleshooting experience are analyzed in this chapter to compile a set of guidelines that designers can follow to reduce the impact these new requirements may have on their time-to-market or to improve their current processes.

5.6.4 Fault Detection and Identification (FDI)

By special modules known as FDI components that operate concurrently with the system, FDI is carried out. Understanding whether a component has failed is the detection task's problem; however, figuring out precisely which failure happened is the identification task's goal. Generally speaking, conditions other than faults may also be subject to detection and identification.

Typically, faults are not directly observable and their occurrence can only be inferred by observing the effects that they have on the observable parts of the system. An FDI component processes sequences of observations (made available by sensors) and triggers a set of alarms in response to the occurrence of faults.

The first ingredient for specifying an FDI requirement is given by the condition to be monitored, called diagnosis condition. The second ingredient is the relation between the diagnosis condition and the raising of an alarm.

An alarm condition is composed of two parts: the diagnosis condition and the delay. The delay relates to the time between the occurrence of the diagnosis condition and the raising of the corresponding alarm. Here are various forms of delay:

- exact (EXACTDEL, after exactly n steps)
- bounded (BOUNDDEL, within n steps)
- finite (FINITEDEL, eventually)

The framework supports further aspects that are important for the specification of FDI requirements.

- The first one is diagnosability, i.e., whether the sensors convey enough information to detect the required conditions. A non-diagnosable system (with respect to a given property) is such that no diagnoser exists that is able to diagnose the property.
- The second aspect is the maximality of the diagnoser, that is, the ability of the diagnoser to raise an alarm as soon as possible and as long as possible, without violating the correctness condition.

The above definition of diagnosability might be stronger than necessary since diagnosability is defined as a global property of the system. In order to deal with non-diagnosable systems, a more fine-grained, local notion of trace diagnosability is introduced, where diagnosability is localized to individual traces.

The formalization encodes properties such as,

- alarm correctness (whenever an alarm is raised by the FDI component, then the associated condition did occur);
- alarm completeness (if an alarm is not raised, then either the associated condition did not occur, or it would have been impossible to detect it, given the available observations).

FDI logic links effects with causes, similar to classical causality theories, but using observables only. An alarm, in this context, corresponds to an effect or, more precisely, to a signal which is triggered by the detection/identification of a given effect. Given a fault F and an alarm A, FDI correctness implies that F is (part of) a cause of A, whereas FDI completeness does not necessarily imply that F is a cause of A since false alarms are possible.

However, correctness and completeness together imply that F is the (unique) cause of A. Finally, diagnosability is related to the realizability of FDI logic, and trace diagnosability corresponds to diagnosability in a specific scenario.

Related to diagnosability, the notion of causality in Fault Tree Analysis (FTA) closely resembles the idea of identifying minimal sets of (necessary and sufficient)

TABLE 5.4
Monitoring Object and Monitoring Purpose for an Electric Vehicle

Order Number	Monitoring Object	Monitoring Purpose
1	Battery voltage	To confirm whether there is a value beyond the range. Low voltage will cause high voltage and insufficient capacity, and will lead to excessive temperatures, gas precipitation, water loss, and battery grid corrosion.
2	Cell voltage	
3	Battery temperature	To identify potential problems and optimize the vehicle operation and cycle life of the battery. Once it exceeds the maximum value, there is a risk of a thermal runaway, necessitating human intervention.
4	Ambient temperature	Battery life will be shortened by an excessively high ambient temperature, and battery capacity will be shortened by an excessively low ambient temperature.
5	Temperature difference	Large temperature difference is because of the inconsistency of the battery, which will cause endurance deterioration.
6	Charge and discharge current	Users should be given access to the battery's health condition information, which can be used to determine the battery's functionality and connection integrity.

causes as in FDI. However, given an effect (Top-Level Event, TLE), FTA is interested in such sets of causes (i.e., faults – identified beforehand) in all possible scenarios, whereas FDI focuses on identifying the causes in a given scenario of interest. FTA has been discussed in Chapter 3.

5.6.5 Case Study: Fault Detection and Identification (FDI) and Diagnosability for an Electric Vehicle

To implement FDI and diagnosability, monitoring object and monitoring purpose for an electric vehicle are summarized in Table 5.4.

5.7 RISK AND BENEFIT: BALANCING THE ENGINEERING EQUATION

Risk and uncertainty are pervasive throughout an engineering product's entire lifecycle. Design defects and manufacturing variability are the most likely causes of product early failures, or infant mortality failures. The interface between engineering design and manufacture is often the weakest link during product development. Late changes made during the design phase in order to achieve better manufacturability may introduce new failure modes and challenge the integrity of the product. The collapse of Hyatt Regency Hotel walkways in 1981, the worst structural failure to date in the United States, was caused by a *simple* design change. The design change that occurred between completion of the original design drawings and the hotel construction effectively doubled the stress on a connection to the walkways.

The catastrophic failure caused by the simple design change resulted in 114 fatalities. Other failure modes causing early structural failures include:

- Failure to meet materials specification;
- Improvisation in construction procedures;
- Unsafe cost-saving choices during product manufacturing;
- Faulty installation.

When a product fails due to an unusually severe environment, such as earthquake, flood, or fire, the failure is independent of product age. The variances with product operational environments often drive the so-called "random" failures that occur after passing through the initial period of infant mortality failures. These failures may happen because the product is used beyond their design-based conditions, the conditions that the engineer judges as highly unlikely or that the engineering team never thought of during product design. Sometimes, these extreme conditions are precipitated by a small incident. On February 15, 1982, a small broken window in the control room of the Ocean Ranger, a mobile offshore drilling rig, eventually caused the ocean vessel to capsize, resulting in the death of every crew member. On that fateful evening, the drilling rig was pummeled by 55-feet-high waves, 80-mile-per-hour winds, and subzero temperatures. The window that broke in the control room was directly to the side of the ballast control panel, which consisted of many electrically operated switches that opened and closed valves on the ballast tanks. The incoming water short-circuited the panel. This not only made keeping the rig floating in a stable manner impossible; the erratic behavior of the ballast control board also caused the rig to lean precariously forward and eventually sink.

Human error frequently enters into system failures; the recorded history in industry can be traced back to the nineteenth-century captains of Mississippi river boats who intentionally blocked safety valves to get more pressure and thus more output from their boilers, thus providing another source of risk and variance during their normal operation.

Even after provisions have been made to minimize the dangers of early failures and random failures, there remains the problem of dealing with the wear-out failures that will become increasingly significant as the product approaches the end of its useful life. Furthermore, the wear-out failures sometimes appear earlier due to ill-conceived design. On June 4, 1979, the roof of the R. Crosby Kempt Jr. Memorial Arena, a structure that had been recognized as one of the nation's finest buildings by the American Institute of Architectures, collapsed when fatigued bolts failed during a rainstorm. These fatigued bolts were made of high-strength steels which perform well under tension and are appropriate for static or constant loading conditions. However, these steels perform poorly under dynamic loading conditions or when structural movement or bending is involved. Investigation following the collapse indicated that the bolts were reduced to between 20% and 25% of their original strength by repeated and frequent rocking of the roof under wind loading during the five-year period that the building had been standing.

Risk and uncertainty are presented everywhere. Careful analysis and sound engineering judgment are needed in connection with risk minimization and risk acceptance.

FIGURE 5.10 What is the risk of NOT accepting a risk?

Why should we accept engineering risk? The answer is that we want to benefit from engineering endeavors. Risk acceptance is an integral part of accepting engineering benefits. Bridges, skyscrapers, airplanes, power facilities, and offshore rigs are sources of risks, but they also provide huge benefits.

As shown in Figure 5.10, risk acceptance is a decision-making process which includes the judgment or assessment about the risk of no action taken – the risk of not accepting a risk. Can we imagine a modem industrial world without bridges, tall buildings, and airplanes?

"Don't accept unnecessary risk" and *"accept risk knowledgeably"* are two very important principles for risk acceptance.

Risk and benefit balance engineering equations and drive the acceptance of engineering risks.

BIBLIOGRAPHY

Ale, B. (1991), "Acceptability Criteria for Risk and Plant Siting in the Netherlands," VIII Meeting 3ASI: Safety Reports in EC, Milan, September 18–19.

Anon. (1980+), Risk Analysis*: An Official Publication of the Society for Risk Analysis*, Plenum Press, New York, NY.

Apostolakis, G., Garrick, B. J. and Okrent, D. (1983), "On Quality, Peer Review, and the Achievement of Consensus in Probabilistic Risk Analysis," *Nuclear Safety*, Vol. 24, No. 6, pp. 792–800.

Chang, S. H., Park, J. Y. and Kim, M. K. (1985), "The Monte-Carlo Method without Sorting for Uncertainty Propagation Analysis in PRA," *Reliability Engineering*, Vol. 10, pp. 233–243.

Committee on Public Engineering Policy. (1972), *Perspectives on Benefit-Risk Decision Making*, National Academy of Engineering, Washington, DC.

Gorden, J. E. (1978), *Structures: Or Why Things Don't Fall Down*, Da Capo Press, New York.

Joksimovich, V. (1985). "PRA: An Evaluation of the State-of-the-Art," *Proceedings of the International Topical Meeting on Probabilistic Safety Methods and Applications*, EPRINP-3912-SR, pp. 156.1–156.10.

Kaplan, S. and Garrick, B. J. (1981), "On the Quantitative Definition of Risk," *Risk Analysis*, Vol. l, No. l, pp. 11–27.

Kyburg, H. E. Jr. and Smokier, H. E. (1964), *Studies in Subjective Probability*, John Wiley & Sons, New York, NY.

Lichtenberg, J. and MacLean, D. (1992), "Is Good News No News?" *The Geneva Papers on Risk and Insurance*, Vol. 17, No. 64, pp. 362–365.

March, J. G. and Shapira, Z. (1987), "Managerial Perspectives on Risk and Risk Taking," *Management Science*, Vol. 33, No. 11, pp. 1404–1418.

Morgan, M. G. and Henrion, M. (1990), *Uncertainty: Guide to Dealing with Uncertainty in Quantitative Risk and Policy Analysis*, Cambridge University Press, Cambridge, England.

Roush, M. L., Modarres, M., Hunt, R. N., Kreps and Pearce, R. (1987), "Integrated Approach Methodology: A Handbook for Power Plant Assessment," SAND87-7138, Sandia National Laboratory.

Rowe, W. D. (1994), "Understanding Uncertainty," *Risk Analysis*, Vol. 14, No. 5, pp. 743–750.

Schlager, N., ed. (1994), "When Technology Fails: Significant Technological Disasters, Accidents, and Failures of the Twentieth Century," Gale Research, Detroit.

Shafer, G. and Pearl, J., eds. (1990), *Readings in Uncertain Reasoning*, Morgan Kaufmann Publishers Inc., San Mateo, CA.

Slovic, P. (1987), "Perception of Risk," *Science*, Vol. 236, pp. 281–285.

Slovic, P. (1993), "Perceived Risk, Trust, and Democracy," *Risk Analysis*, Vol. 13, No. 6, pp. 675–682.

Wahlstrom, B. (1994), "Models, Modeling and Modellers: An Application to Risk Analysis," *European Journal of Operations Research*, Vol. 75, No. 3, pp. 477–487.

Wang, J. X. (1991), "Fault Tree Diagnosis Based on Shannon Entropy," *Reliability Engineering and System Safety*, Vol. 34, pp. 143–167.

Wang, J. X. (1996), "Complexity as a Measure of the Difficulty of System Diagnosis," *International Journal of General Systems*, Vol. 24, No. 3, pp. 257–269.

Wang, J. X. (2002), *What Every Engineer Should Know about Decision Making Under Uncertainty*, CRC Press, Boca Raton, FL.

Wang, J. X. (2017), *Industrial Design Engineering: Inventive Problem Solving*, CRC Press, Boca Raton, FL.

Wang, J. X. (2019), "Complexity as a Measure of the Difficulty of System Diagnosis in Next Generation Aircraft Health Monitoring System," SAE Technical Paper 2019-01-1357, doi:10.4271/2019-01-1357.

Wang, J. X. (2019), "A Dynamic Fault Tree Approach for Time-Dependent Logical Modeling of Autonomous Flight Systems," SAE Technical Paper 2019-01-1358, doi:10.4271/2019-01-1358.

Wenk, E. et al. (1971). *Perspectives on Benefit-Risk Decision Making*, The National Academy of Engineering, Washington, DC.

Zadeh, L. A. and Kacprzyk, J., eds. (1992), *Fuzzy Logic for the Management of Uncertainty*, John Wiley and Sons, New York, NY.

6 From Risk Engineering to Risk Management

6.1 PANAMA CANAL: RECOGNIZING AND MANAGING RISK

As in shown in Figure 6.1a, the Panama Canal is a 50-mile long waterway that bridges the Pacific and Atlantic Oceans. Begun in the 1880s by French engineers and completed in 1913 by the U.S. engineers, the canal proved to be one of the greatest engineering feats of the ages. Panama possessed one of the most difficult and deadliest terrains on earth, causing numerous complications. The huge engineering project ultimately took two nations – France and the United States, the equivalent of 7 billion dollars, and 25,600 lives.

Early American traffic developed along waterways rather than overland. Rivers were man's first highways, but were inconveniently punctuated with rapids and falls. In order for any ship to pass from the Atlantic Ocean to the Pacific Ocean, or vice versa, one would have to either go around the tip of South America (the Horn) or go around the most northern part of North America (north west passage does not exist because of ice all year). This was very time-consuming and extremely inefficient.

6.1.1 Risk Recognition at an Earlier Stage

In 1878, Ferdinand de Lesseps, the builder of the Suez Canal, announced that he would build a sea-level canal across Panama. Why did he pick Panama? For starters, Panama was in Central America, perfectly located, and was only 40 miles from shore to shore. Panama was a crossroad of global trade. Creating a canal in Panama would complete a water circle around the world. Time and mileage would be cut down dramatically when traveling from the Atlantic to the Pacific or vice versa. It would save a total of 18,000 miles on a trip from New York to San Francisco.

Concept design is a critical part of engineering projects. Although several of the delegates of the International Congress strongly recommended a multi-tiered canal with locks, de Lesseps insisted on a sea-level canal through Panama. With the success he had building the Suez Canal in Egypt just ten years earlier, de Lesseps was confident that he would complete the water circle through the world. He believed that if a sea-level canal worked when constructing the Suez Canal, it most certainly must work for the Panama Canal. But he had underestimated the risk of this challenging project.

The excavation of the cut through the continental divide presented a major technical risk. The French workers cut away the jungles and trees by hand. Much of the work was done quickly at first, but as the days dragged on and the cuts became deeper, the French were in for disaster. As the workers were excavating the dirt to create the deep cut for the canal, the amount of excess soil became enormous. The

DOI: 10.1201/9781003371014-6

FIGURE 6.1a Panama Canal: Gateway to the world.

French had hauled the soil away from the excavation sites by trains of little dump cars to some adjacent valley where the soil was then dumped and allowed to build up. This was a quick and economical solution to the problem, but caused much of the problems the French were beginning to encounter. When rain fell, the piles of extra soil slipped and created mud slides, destroying the railroads used to dump the dirt, thereby breaking down their whole system.

French engineers could not stop mud and rock slides from destroying newly dug cuts. Repeated landslides had made it clear that substantially greater volumes of earth and rock had to be excavated to achieve stable slopes through the isthmus; de Lesseps had made some fatal mistakes. He was wrapped up in the idea that a sea-level passageway across Panama would work best. Worst of all, he failed in waging all-out warfare against malaria and yellow fever. Many lives were claimed by these deadly diseases. For the ten years that they had occupied Panama, 20,000 French workers and engineers had died. During two of the French's worst wet seasons, the average daily burial rate was 35, due to hot, humid, and extremely damp conditions with torrential downpours.

After ten years and 20,000 dead, the French finally pulled out of Panama, but only due to lack of money. The French company was in financial trouble and unable to raise badly needed capital. Being one of the biggest financial losses in history, the French completed only one-third of the work. The canal company ended up in receivership. All that the French left behind was a partially dug up canal and a cemetery outside of Panama City, as a silent witness to the risk of the engineering project: casualty, cost overrun, and schedule slip.

6.1.2 Risk Management at a Later Stage

After the title to the canal property was transferred from the French company to the United States on April 22, 1904, John Stevens, the greatest railroad engineer of the time, took over as the chief engineer. He knew where the French had gone wrong and wanted to correct these mistakes. The first step on the agenda was to clean up the entire city. Getting rid of ill-health conditions and making for a liveable environment was one key to success. Stevens stopped all work on the canal in order to make conditions fit to live in. He granted William Gorgus all the money he needed to rid the zone of pestilence of any kind and eradicate the disease of yellow fever.

The project again was at the concept design stage, whose importance was emphasized by the project delay during the first year of American construction. Stevens envisioned a canal not at sea level, but one with locks and gates to raise ships up and over Panama. Miter gates, also known as lock gates, are the tools that help to contain the water in each separate chamber of a canal. Each gate is connected to the walls with hinges and swings open like double doors. These gates meet at an angle pointing toward the flow of the water making them self-securing. When there is a difference in the height of the water between two adjacent chambers, the pressure holding the gates is at its greatest. The water is doing the work to hold the gates closed so that extra energy is not required. The most impressive aspect of the miter gates constructed for the Panama Canal is their immense size. They are up to 82 feet high and 65 feet wide, the width of the canal being 110 feet. Due to the extreme variation of the Pacific tides, the highest and heaviest gates, 745 tons, are the lower locks located at Miraflores.

The canal designers had to deal with the risk of failing to maintain enough water for large ships as well as the risk of loss of power supply. The vital factor in the whole plan of the construction of the locks is water. Water lifts or lowers ships and makes the tremendous hollow locks, which are three times heavier than those ever built, virtually weightless. What makes this canal so amazing is its self-sufficiency. Due to gravity, the great dam of the Gatun Spillway is able to generate enough electrical current to run all the motors to operate the canal as well as the locomotives in charge of towing the ships through the canal. There is no force required to adjust the water level between the locks aside from the force of gravity. As the locks operate, the water simply flows into the locks from the lakes or flows out into the sea-level channels.

Quality control and fault-tolerant design were used during the construction of the canal. When building the locks out of steel gates, inspectors went down inside the gates through several manholes to check the rivets that connected the steel plates to a grid of steel girders. All imperfect rivets were cut out and replaced. The watertightness of each gate was tested by filling the gate leaves with water. Another way the engineers made sure of the safety of the canal was to duplicate all of the gates. Each set of double doors was backed by another to ensure that if one set failed to work correctly or was damaged by a ship, another set would be able to assume the proposed responsibility. This added to a total of 46 gates in the Panama Canal.

Still the excavation of the Culebra Cut through the continental divide presented the most compelling technical risk. The French never realized that digging the cut was more a problem of moving the excess soil out of the way than the actual excavation. This is where the American engineers took a step forward instead of several steps back as did the French. Using his railroad expertise, Stevens devised a system of railroads to haul dirt away, thereby increasing efficiency of excavation. The dirt removed was hauled away and used to dam the Chagras River and create a manmade lake. As years pass, the soil will continue to settle increasing the strength of this great dam. This solved many of the problems: the soil was being hauled away effectively, the dam was being built at the same time to administer the needs of the locks by creating the lake and the essential water supply, and lastly, reusing and recycling was being practiced, saving time, money, and most importantly, our earth resources. This is the perfect example of engineering design in combination with the use of natural resources.

After Stevens quit unexpectedly, Colonel George Washington Goethals was appointed to the job. The central division, which included Gatun Lake and the infamous Culebra Cut, came under the direction of Major David Du Bose Gaillard. Gaillard died in 1913 but the engineer's name remained with the engineering project. Changing the name of Culebra Cut to Gaillard Cut recognized the engineer's central role for removing 219 million cubic yards of earth during the final seven years of the project. The construction of the canal finally concluded on May 31, 1913. Today, the canal is still one of the most highly traveled waterways in the world, handling 12,000 ships per year. The 50-mile crossing takes about nine hours to complete, an immense time savings as compared with rounding the tip of South America.

The Panama Canal is one of our world's greatest engineering feats. Much work, thought, knowledge, and experience went into the completion of this engineering wonder. Without the contribution of both the American engineers and French engineers, this canal would not be as successful as it is today. Through the failure of French engineers at the earlier stage, American engineers recognized the risk behind the huge engineering project. Without the mistakes made by the French engineers, the American engineers would not have learned and therefore would not have been able to make the necessary corrections leading to their success in risk management.

6.2 PROJECT RISK ASSESSMENT: QUANTIFY RISK TRIANGLE

Risk management has been defined as the process of balancing risk with cost, schedule, and other programmatic considerations (Figure 6.1b). It consists of risk identification, risk assessment, decision-making on the disposition of risk, and tracking the effectiveness of the results of the actions resulting from the decisions. It includes maximizing the results of positive events and minimizing the consequences of adverse events. Project risk management extends beyond ordinary project planning and control with regard to assessing, reducing, and controlling the project risks. It becomes increasingly important as customers are less tolerant than ever before of any defects.

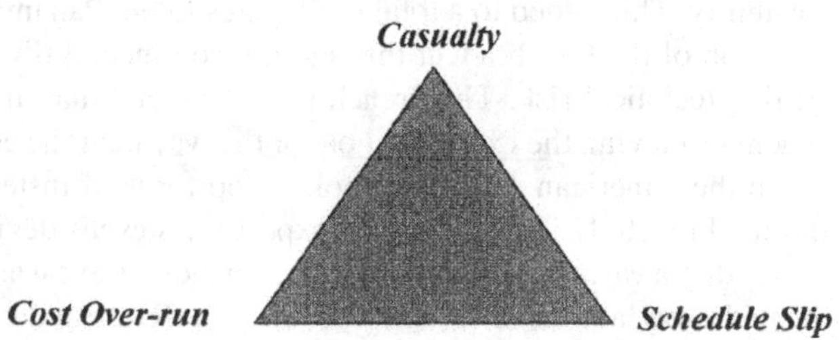

FIGURE 6.1b Risk triangle with engineering projects.

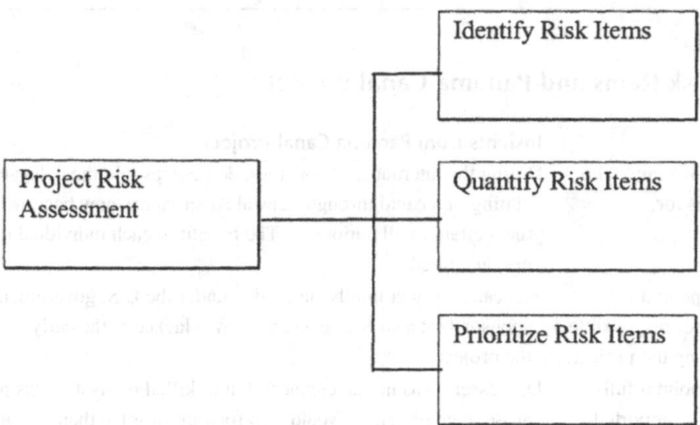

FIGURE 6.2 Elements of project risk assessment.

Risk management includes both qualitative and quantitative factors. There are two stages in the process of project risk management: risk assessment and risk control. These processes interact with each other and with the processes in the other knowledge areas as well. For example, risk control cannot be effective without a previous risk assessment. As shown in Figure 6.2, project risk assessment has three elements: identify risk items, quantify risks, and prioritize risks.

6.2.1 Identify Risk Items

A risk item is any uncertainty with technical performance, resource, or schedule outcome. Risk identification consists of determining which risk items are likely to affect the project and documenting the characteristics of each. For example, the French engineers underestimated the likely duration of the excavation of the cut through the continental divide during the Panama Canal project. This risk item had a very significant impact on the project schedule. Some sample risk items and insights from the Panama Canal project are summarized in Table 6.1. Risk item identification is not a one-time event; it should be performed on a regular basis throughout the project.

Risk item identification should address both internal and external risks. Internal risks are things that the project team can control or influence, such as resource management and cost estimates. External risks are things beyond the control or influence of the project team, such as market shifts or government actions.

Risk item identification may be accomplished by identifying cause-consequence relationships (what could happen and what will induce) and consequence-cause relationships (what outcomes are to be avoided and how each might occur). The former is the event tree analysis, while the latter is the fault tree analysis given in Chapter 3.

6.2.2 Quantify Risk Items

Risk quantification involves evaluating risks and risk interactions and assessing how those areas of uncertainty can impact the performance of a project, either in

TABLE 6.1
Sample Risk Items and Panama Canal Project

Risk Items	Insights from Panama Canal Project
Failure to understand who the project is for;	During the international congress, de Lesseps told the delegations: "The cutting of a canal through Central America has now become essential to the welfare of all nations...". The benefit to each individual nation was not quantified.
Failure to appoint an executive user responsible for sponsoring the project;	The canal project finally succeeded under the U.S. government sponsorship; a strong sponsorship was lacked at the early stage of the project.
Failure to appoint a fully qualified and supported project manager;	De Lesseps was not an engineer, but a skilled motivator: "as problems arose, men of genius would step forward to solve them... science would find a way". A systematic project management approach apparently did not exist here.
Failure to define the objectives of the project;	"The building we are met with is dedicated to science, and the impartial serenity of science will impress itself upon your deliberations", said de Lesseps. This is an extremely vague statement for such a huge engineering project.
Failure to secure commitments from people who are needed to assist with the project;	The French finally pulled out of Panama due to a lack of necessary financial support.
Failure to estimate technical, cost, and schedule risk accurately;	De Lesseps had underestimated the task's magnitude, which greatly exceeded the Suez Canal he accomplished in Egypt ten years earlier.
Design errors;	De Lesseps insisted on a sea-level canal design and refused to consider a canal containing locks.
Failure to provide a good working environment for the project;	One reporter, Sheldon B. Liss, described Panama during French construction as "a foul hole, by comparison, the ghettos of White Russia, the slums of Toulon, Naples, and old Stamboul...deserve prizes for cleanliness. There are neither sewers nor street cleaners,...toilets are quite unknown, all the rubbish is thrown into the swamps or into rubbish heaps. Toads splash in the liquid muck..., rats infest the solid filth..., snakes hunt both toads and rats; clouds of mosquitoes swarm into home".

duration, cost, or meeting the users' requirements. It is primarily concerned with determining which risk items warrant risk control.

Figure 6.3 plots the probability of occurrence of a risk, which is another way of saying how uncertain the success of the task would be, against the impact. By impact, we mean the severity of the effect on either the budget, the timeliness of project completion, or the ability of the project to meet the users' requirements. Whether the severity of impact or the probability is high or low is a matter of the judgment of the risk assessor and the project manager.

We have classified four sectors of the graph, perhaps whimsically, as follows:

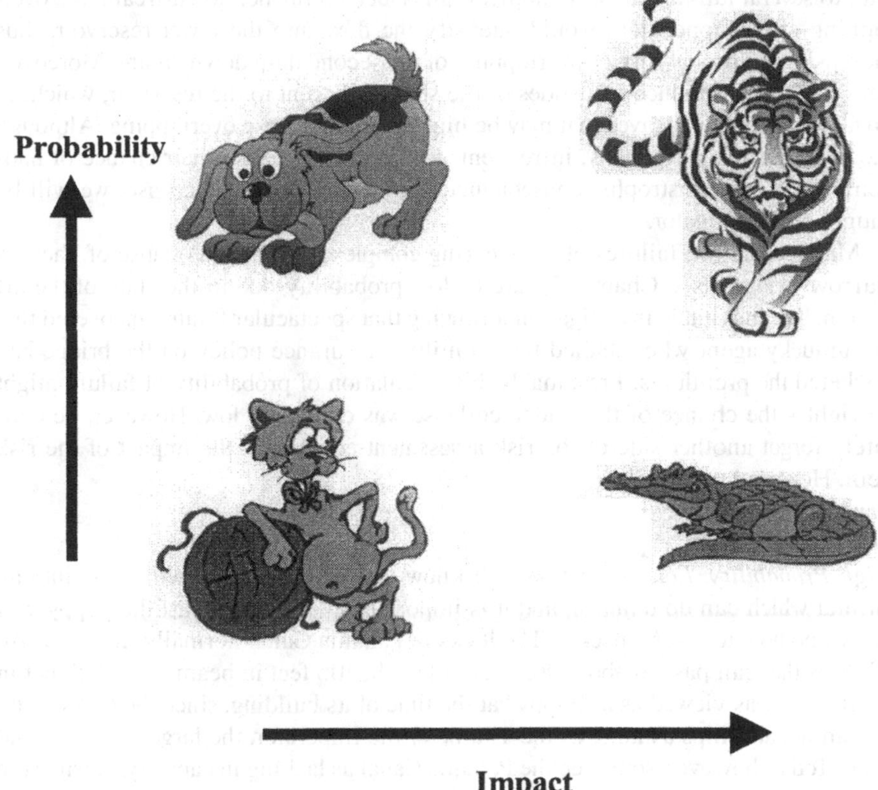

FIGURE 6.3 Classifying risks.

6.2.2.1 Tigers

High Probability, High Impact. These are dangerous animals and must be neutralized as soon as possible. The illnesses such as yellow fever were apparently a tiger claiming the lives of thousands of workers during French construction. When the Americans fully began their effort in 1904, the chief sanitary officer was William Crawford Gorgas, who had experienced much success in eradicating yellow fever and malaria in Cuba. After John Stevens took over as the chief engineer, his first step on the agenda was to clean up the entire city. Stevens stopped all work on the canal in order to make conditions fit to live in. To eradicate the disease of yellow fever, he granted William Gorgus all the money he needed to neutralize the tiger for the project.

6.2.2.2 Alligators

Low Probability, High Impact. These are dangerous animals which can be avoided with care. However, we all remember the old joke that it is difficult to remember when one is up to the ears in alligators that the original objective was to drain the swamp. The overtopping of a dam resulting from a heavy rainstorm over the watershed could lead to several failure paths. If another dam is located further downstream, the overtopping of the upper dam would intensify the flow into the lower reservoir, thus increasing the probability of overtopping of the second dam downstream. Moreover, heavy rain could induce landslides of the slopes adjacent to the reservoir, which, in turn, could generate waves that may be high enough to cause overtopping. Although those events are, fortunately, infrequent, we should not forget their chance of happening and the catastrophic consequences they could cause; otherwise, we will be caught by the alligator.

Major structural failures of engineering complexes, like the collapse of Tacoma Narrow Bridges (see Chapter 5), are of low probability due to the state-of-the-art design. The inevitable investigation following that spectacular failure uncovered that the unlucky agent who handled the $6 million insurance policy on the bridge had pocketed the premiums. Presumably, his calculation of probability of failure might be right – the chance of the bridge collapse was extremely low. However, he definitely forgot another side of the risk assessment equation – the impact of the risk item. He ended up in jail.

6.2.2.3 Puppies

High Probability, Low Impact. We all know that delightful pup will grow into an animal which can do damage, and it is important to know how fast the puppy will grow and how to re-engineer it. The locks in Panama Canal normally limit the size of ships that can pass to about 965 feet in length, 106 feet in beam, and 39.5 feet in draft. This was viewed as a "puppy" at the time of its building, since the locks could accommodate ships as large as the Titanic or the Imperator, the largest ships at that time. Today, however, some see the Panama Canal as lacking in capacity. A tripartite committee, comprising representatives of the United States, Japan, and Panama, is presently considering alternatives for improving the canal.

6.2.2.4 Kittens

Low Probability, Low Impact. The largest cat is rarely the source of trouble, but, however, a lot of effort can be wasted on assessing or training it.

List each of your identified risks, decide on the probability of occurrence of each, and define the expected impact on schedule, budget, and ability to meet the users' requirements.

6.2.3 Prioritize Risk Items

Risk item prioritization establishes which risk items should be eliminated completely, because of potential extreme impact, which should have regular management attention, and which are sufficiently minor to avoid detailed management attention. Tigers must be neutralized, i.e., the risks must be mitigated early on. Alligators must be watched and there must be an action plan in place to stop them from interfering with the project. Puppies similarly must be watched, but less stringently and with less urgent containment plans. Kittens can be ignored at the peril of the project manager. Risk item prioritization is complicated by a number of factors, including, but not limited to:

Opportunities and threats can interact in unanticipated ways. For example, the schedule delays during the first year of America's Panama Canal construction provided an opportunity for reconsidering a multi-tiered canal with locks, which reduced the overall project duration.

A single risk item can cause multiple consequences. For example, the repeated landslides during earlier Panama Canal construction caused heavy causality, cost overruns, and schedule delays.

6.3 WHY ARTIFICIAL INTELLIGENCE PROJECTS FAIL – HOW TO AVOID?

With hundreds of use cases, artificial intelligence (AI) and machine learning (ML) are revolutionizing numerous sectors and business practices. But for businesses, creating and successfully deploying an AI project to business processes present major difficulties, which is inhibiting the organization's adoption of AI.

6.3.1 Uncertain Business Goals

Although AI is a powerful technology, its implementation alone won't lead to success unless it is used to solve a specific business problem and achieve certain business objectives. Companies must first identify and define business problems before deciding whether AI approaches and tools will be helpful in solving them. This is preferable than starting from the answer for an unsolvable business problem.

Additionally, it is difficult to estimate the costs and possible advantages of an AI project because:

1. Building and training AI models are experimental processes that could take a lot of time and trial-and-error.
2. AI models attempt to solve probabilistic business challenges; the results may vary depending on the application scenario.

A clearly stated business objective can help determine whether AI is the best tool or whether there are other tools or methodologies that can be used to solve the challenge at hand. This can help businesses avoid paying unneeded expenses.

Because they lack well-defined business objectives, AI projects frequently never get off the ground. The typical organization finds it difficult to focus on a quantifiable business goal before creating a tool or system to address an issue. What this implies is that projects involving AI can begin as creative, flashy ideas unrelated to the core economic value drivers. Before beginning an AI project, the following queries should be raised:

1. What specific business issue am I attempting to resolve?
2. How will the AI be applied to effectively address this business issue?

If you are unable to respond to these inquiries, your project probably lacks a distinct commercial goal. In that scenario, you ought to cease writing code. Spend your efforts correctly defining a clear purpose rather than wasting time on a project that is unlikely to help the company enhance its revenue.

6.3.2 Data of Poor Quality

The most important resource for any AI effort is data. To guarantee the availability, quality, integrity, and security of the data they will utilize in their project, businesses need to design a data governance strategy. Working with stale, incomplete, or biased data can result in garbage-in-garbage-out scenarios, project failure, and resource waste.

The effectiveness of the deep learning algorithms and AI technologies used to combat the COVID-19 epidemic is a good illustration of the use of high-quality data in AI initiatives. Researchers examined hundreds of AI technologies created for COVID-19 diagnosis or risk prediction of patients from data such as medical imaging and came to the conclusion that none of them are appropriate for practical applications. The majority of the difficulties were related to poor data quality, such as unknown sources and incorrect labeling:

1. As samples of non-COVID cases, many models employed a dataset of child chest scans that were in good health. In the end, rather than COVID examples, the AI learned to recognize children.
2. In several instances, the AI made advantage of hospital-specific text typefaces to classify the scans as indicators of COVID risk.
3. Other times, chest scans were taken when the patient was lying down for some and standing for other patients, using ML models. The AI learns to anticipate patients' risk based on their positions because a patient who is lying down is more likely to be ill.

Before starting an AI project, businesses should make sure they have enough relevant data that accurately represents their company processes, has the right labels, and is appropriate for the AI tool being used. Otherwise, using AI technologies for decision-making can lead to inaccurate results and be risky.

The chances of an AI project might be quickly destroyed by poor data management. Sanitized, properly labeled, and documented data are required. What this implies is that early implementation of a standard will guarantee efficient data collection, archival, and management. This contains elements that give the data its "meaning" and enable model training. These procedures ought to be established at the beginning of a project and ought to remain constant over time. Your team is in an excellent position to create a successful AI project if it can provide answers to these questions. Although you might still encounter obstacles or problems with data management, having these procedures in place will help to ensure that you stay clear of the dangers.

6.3.3 Teams Not Working Together as a Team

An AI team working alone on an AI project is not the best way to ensure success. AI scientists, AI engineers, IT experts, designers, and line of business personnel must work together to create a successful AI project. Companies would benefit from creating a cooperative technical environment if they could:

1. Make sure the AI project's output is adequately integrated into their company's broader technical architecture.
2. Streamline the process of developing AI.
3. Exchange knowledge and expertise, and create best practices.
4. Deploy AI solutions at scale.

To bridge the gap between various teams and scale the operationalization of AI systems, there are sets of approaches called as DataOps and MLOps. Furthermore, better collaboration can be achieved by creating a federated AI Center of Excellence (CoE) where data scientists from many business disciplines can cooperate.

6.3.4 Inadequate Talent

The shortage of qualified data science specialists, per a survey, is the main obstacle to the adoption of AI for organizations. Due to this skill scarcity, building a strong data science team can be expensive and time-consuming. Companies shouldn't expect their AI program to provide significant results if they don't have a team with the right training and business domain experience.

Businesses must weigh the advantages and disadvantages of forming internal data science teams. Check out our articles on in-house AI and ML outsourcing to learn more about the benefits and drawbacks of creating in-house teams vs. outsourcing. Outsourcing may initially be a more affordable option to implement AI applications, depending on your business goals and the size of your operations.

The appropriate individuals must be involved from the beginning for a project to be successfully completed. This refers to both domain and AI knowledge in the context of AI. What this means is that the proper people must be involved in an AI project – or any project, for that matter – from the very beginning. By using internal and external data sources, this will necessitate finding, hiring, and retaining highly qualified personnel with backgrounds in both ML (for creating models) and subject matter expertise (for deploying models).

6.3.5 Absence of Standards and Governance

If the right standards and governance aren't in place, AI initiatives may also come to a dead end. These must be established in the "early days" of the project to prevent risks like misconfiguration, security flaws, or incompatibility from building up. This means that in order to avoid being doomed from the start due to procedural obstacles, AI initiatives must establish governance and standards early on. It is crucial to stress again that one of the drawbacks of AI is its inexplicability. An algorithm's decisions must be "explained" to a business line owner in a way that they can comprehend and accept. Setting up the proper governance and standards from the beginning will be necessary for this.

6.3.6 Lack of Ownership and Commitment from the Leadership

This is a common mistake in all projects, not just those involving AI. The initiative won't have the resources necessary for success without the dedication and ownership of the leadership. What this implies is that it is unlikely that any major work can get done on an AI project without skilled expertise being accessible or committed. An AI project can only be successful if it has capable leaders who are dedicated to it.

6.3.7 QA Testing Not Being Used

QA testing is frequently not integrated at all phases of the product development process by businesses in all sectors. It is incorrectly regarded as an add-on, as a formality to ensure that a product functions well, rather than as a tool that can be used to optimize the product iteratively.

The fact that this mindset toward QA testing is unworkable in light of the realities of AI development is one reason why AI/ML initiatives fail. Contrary to typical software development, faults cannot be rectified by a simple software update; rather, QA testing failures can only be corrected by starting over from scratch. If your AI is not performing as expected, there was either an issue with the training data or the training data caused the model to be skewed in the wrong way. In any case, this calls for returning to step one and gathering fresh artifacts from the training data.

Businesses which don't incorporate outcome evaluation at every stage of the AI development process add extra work to their own workload. Companies need to train and test more iteratively rather than training the algorithm with a single massive dataset and then putting the AI to the test. Agile testing that is "built in" will assist reduce wasteful spending, accelerate schedules, and enable more effective resource allocation.

6.3.8 How to Ensure Successful AI/ML Projects?

The availability of technologies like AI as software is revolutionizing private, public, and business lives. AI is the future. Keep in mind and steer clear of the aforementioned common causes of AI project failure in order for the project management underlying all of these revolutionary advancements to go successfully.

Companies must take a programmatic approach to creating AI. Companies should think about combining each stage of the process as a part of a comprehensive program rather than considering each stage of the process as a separate project. AI development is an iterative, agile process in which teams must collaborate rather than operate in isolation. Each team is led by a program leader who is responsible for the program's success.

6.4 PROJECT RISK CONTROL

Most people tend to think that by having performed a risk assessment, they have done all that is needed. Far too many projects spend a great deal of effort on risk assessment and then ignore risk control completely. Similar to risk assessment, project risk

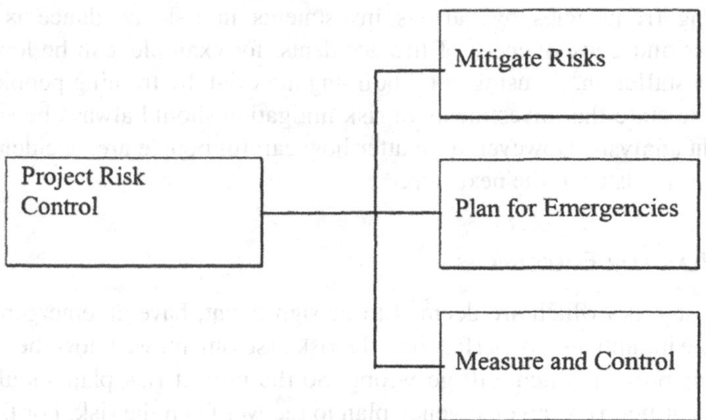

FIGURE 6.4 Elements of project risk control.

control has three elements: mitigate risks, plan for emergencies, and measure and control residual risks (see Figure 6.4).

6.4.1 Mitigate Risks

Risk mitigation takes whatever actions are possible in advance to reduce the effect of risk. It is better to spend money on mitigation than to include contingency in the plan. You would do this for all those risks categorized above as Tigers. We can mitigate risks by reducing either the probability or the impact. Remember that we identified the risk by seeking uncertainty in the project. The probability can be reduced by actions upfront to ensure that a particular risk is reduced. As a risk mitigation feature in the Panama Canal, the leaves of the gates were designed to close in a V, with the V always pointing upstream against the pressure of the water. Only when the water pressure on both sides of the gates was equal, that is, level of the water was the same on both sides of the gates, could the gates be structured at each end of the uppermost lock that was identical to the set of main gates. These gates would be used in case the main gates did not work properly or were rammed by an out-of-control ship.

Risk reduction consists of reducing uncertainties, reducing consequences, avoiding risks, and transferring risks. Although there is always a margin of uncertainty associated with the results of a quantitative risk assessment, experience has shown that the envelope of uncertainty is often swamped by the much larger variation in the effectiveness of risk mitigation measures themselves. Thus, the benefit to cost ratio is generally a useful discriminator between upgrade options because some options are usually found to be clearly worthwhile, and others clearly ineffective, even allowing for uncertainty in the risk and cost estimates.

Risk transfer is a common means for risk control. Risk transfer refers to transferring a risk to someone else through, for example, a contract. The risk for transport damages is a typical example where it is possible to assign the risk to the buyer through a contract wording.

Reducing frequencies by various investments in risk avoidance is common. Frequencies and consequences of fire accidents, for example, can be lowered by a sprinkler installation, by using good building material, by training people, etc. It is important to state that investments in risk mitigation should always be subject to a cost benefit analysis. However, no matter how careful people are, accidents do happen and this leads us to the next step.

6.4.2 Plan for Emergencies

For all those risks which are deemed to be significant, have an emergency plan in place before it happens. By performing the risk assessment, we know the most likely areas of the project which will go wrong. So the project risk plan should include, for each identified risk, an emergency plan to recover from the risk. For the Panama Canal, the emergency planning in the locks was the keynote. There was provided above each lock an emergency dam, which can be closed in case of accidents. These emergency dams functioned like a movable bridge that could swing out over the lock and lower gates quickly to cut off any unexpected flow of water. Another emergency planning measure was a 12-ton iron fender chain that ran between the chamber walls. If the ship was proceeding normally, the chain was lowered into a special groove in the chamber floor. In the remote chance that the ship was out of control, an automatic mechanism would raise the chain until the ship was stopped.

A risk that has been mitigated may still be a significant and dangerous risk – it is rare for a tiger to be converted to a kitten by action before the event. These will require emergency plans as well for alligators and puppies. Kittens can probably be allowed to play at will, provided that we are satisfied they really are kittens.

Good contingency plans, emergency plans, are essential when a major loss has happened or is in the process of escalation. Trained personal and practiced procedures can work wonders.

6.4.3 Measure and Control Residual Risks

Track the effects of the risks identified and manage them to a successful conclusion. The owner of each risk should be responsible to the project manager to monitor his risk and to take appropriate action to prevent it from happening or to take recovery action if the problem does occur.

Nothing can be controlled that cannot be measured. In a project, there are three things that can always be measured – the schedule, the cost, and the users satisfaction. Note that the latter is not the same as whether or not the project meets the original specifications. If the project meets all three criteria, it is right to consider it a successful project. Let's now measure the Panama Canal project by these three criteria.

The cost of the canal must be figured in several ways. In dollars, the total price is staggering. The Americans spent $352 million. Add in the French expenditures, and the total peaks out at approximately $639 million. In 1914, this made the Panama Canal the greatest single construction project in American history. Another cost must be included, however: the cost in lives. According to American records beginning in 1904, 5,609 workers died from diseases and accidents. Add to this the number of

French lives, and the number swells to nearly 265,000. This represents the magnitude of the risk with this challenging project.

Beyond the risk assessment numbers, there are other impressive statistics. The Panama Canal was completed ahead of schedule by six months, despite delays caused by landslides. The project was finished under budget, as well, by about 23 million dollars.

Insurance or an insurance-like arrangement such as bonding is often available to residual risks. As the output of the overall risk assessment and risk control processes, a risk management plan should be documented including the procedures that will be used to manage risk throughout the project. In addition to documenting the results of the risk assessment and risk control plan, it should cover who is responsible for managing various areas of risk, how the initial risk assessment outputs will be maintained, how contingency plans will be implemented, and how reserves will be allocated.

The essence of risk management is the avoidance of anything which extends the schedule, increases the costs, or impairs the users' satisfaction with the product of the project. As summarized in Table 6.2, risk management means anticipating,

TABLE 6.2
Summary of Project Management Activities

	Project Priority			
Area	4	3	2	1
	Project cost is not significant; familiar project tasks; low risk technologies; low impact of project failure.	*Project cost is moderate; some unfamiliar tasks and technologies; medium risk impacts.*	*Project cost is significant; unfamiliar project tasks; new or innovative technologies.*	*Complex project with potentially high risk issues and exposures; uncertainties inherent in plan and technologies.*
Risk Assessment	Use management judgment to list expected risk areas; compare project objectives to risk items and identify manageable risks.	Document risk areas and evaluate low-medium-high risks; identify risks with significant impact.	Establish structured methodology for identifying, quantifying, and assessing all potential project risks.	Document risk identification, probability, and consequences for objectives, specifications, and stakeholder interests; conduct extensive quantitative risk assessment including uncertainty analysis.
Risk Mitigation	Identify technologies or approaches presenting unattractive risks; plan actions to minimize risk exposure.	Assign study teams to develop risk avoidance and/or mitigation plans for excessive risk items.	Develop risk deflection strategies for all significant project risks; incorporate contingency actions into project plans.	Conduct cost/benefit analysis to select candidates and strategies for risk reduction.

(Continued)

TABLE 6.2 *(Continued)*
Summary of Project Management Activities

	Project Priority			
Area	4	3	2	1
	Project cost is not significant; familiar project tasks; low risk technologies; low impact of project failure.	*Project cost is moderate; some unfamiliar tasks and technologies; medium risk impacts.*	*Project cost is significant; unfamiliar project tasks; new or innovative technologies.*	*Complex project with potentially high risk issues and exposures; uncertainties inherent in plan and technologies.*
Risk Management Plan	Structure project management approach and work processes specifically to address risk areas and exploit opportunities; use modular and phased approaches to compartmentalize risk and minimize risk compounding.	Address each significant risk item and apply a specific project management or technical approach to minimize, manage, and control risk events.	Develop action steps and staffing to reduce uncertainties and control risk areas.	Document plans for risk focused management attention to respond to all risk areas; apply risk management best practices (or lessons learned) to project life cycle.
Risk Metrics	Identify issues to monitor risk items using subjective or qualitative risk assessments; follow-up reporting risk issues at periodic reviews; highlight high risk areas and adverse trends.	Assign all risk areas low-medium-high assessment and update and report status and trends; use qualitative or subjective metrics if none better available.	Develop measurable indicators of risk exposure; report status and trends.	Develop metrics for risk areas, report status and trends; track impact of risk control actions on lessening risks; focus on areas that corrective actions are still not closed.

before the start of the project, unexpected situations beyond the project manager's full control that can undermine the success of the project. Planning ahead for these unexpected situations includes:

What is likely to happen to this project? (Risk item Identification)
Which risks have probability and impact for concern? (Quantify and prioritize risk items)
What can be done to manage the threats – avoid, mitigate, contingency? (Risk mitigation and emergency planning)
How might the risk plan change for each phase of the project? (Measure and control residual risks)

Risk management is about expecting the unexpected problems. By anticipating these unexpected events, problems, and conditions, and modifying plans appropriately, these situations can be avoided, minimized, or managed without upsetting the project commitments. Managing these contingencies upfront eliminates the costly consequences to meeting objectives – technical performance, budgets, and schedule – as well as the effects on the plans and interests of project stakeholders.

BIBLIOGRAPHY

Augustine, Norman. (1994), "Is Any Risk Acceptable Today?," *Across the Board*, Vol. 31, No. 5, pp. 14–15.

Bryant, Michael W. et al. (1992), "Risk Management Roundtable: Improving Performance with Process Analysis," *Risk Management*, Vol. 39, No. 11, pp. 47–53.

Burlando, Tony. (1994), "Chaos and Risk Management," *Risk Management*, Vol. 41, No. 4, pp. 54–61.

Chicken, John C. (1994), *Managing Risks and Decisions in Major Projects*, Chapman & Hall, London.

Cooper, Dale F. (1987), *Risk Analysis for Large Projects: Models, Methods, and Cases*, Wiley, New York.

Defense Systems Management College. (1983), *Risk Assessment Techniques: A Handbook for Program Management Personnel*, DSMC, Ft. Belvoir.

Englehart, Joanne P. (1994), "A Historical Look at Risk Management," *Risk Management*, Vol. 41, No. 3, pp. 65–71.

Esenberg, Robert W. (1992), "Risk Management in the Public Sector," *Risk Management*, Vol. 39, No. 3, pp. 72–78.

Grose, Vernon L. (1987), *Managing Risk: Systematic Loss Prevention for Executives*, Prentice-Hall, Englewood Cliffs, NJ.

Kurland, Orim M. (1993), "The New Frontier of Aerospace Risks," *Risk Management*, Vol. 40, No. 1, pp. 33–39.

Lewis, H.W. (1990), *Technological Risk*, Norton, New York.

Lundgren, Regina. (1994), *Risk Communication: A Handbook for Communicating Environmental, Safety and Health Risks*, Battelle Press, Columbus, OH.

McKim, Robert A. (1992), "Risk Management: Back to Basics," *Cost Engineering*, Vol. 34, No. 12, pp. 7–12.

Moore, Robert H. (1992), "Ethics and Risk Management," *Risk Management*, Vol. 39, No. 3, pp. 85–92.

Moss, Vicki. (1992), "Aviation & Risk Management," *Risk Management*, Vol. 39, No. 7, pp. 10–18.

Petroski, Henry. (1994), *Design Paradigms: Case Histories of Error & Judgement in Engineering*, Cambridge University Press, Cambridge.

Raftery, John. (1993), *Risk Analysis in Project Management*, Routledge, Chapman and Hall, London.

Risk Management Concepts: Risk Management Seminar. (1993), NASA Headquarters, Washington, DC, March 17.

Schimrock, H. (1991), "Risk Management at ESA," *ESA Bulletin*, Vol. 67, pp. 95–98.

Sells, Bill. (1994), "What Asbestos Taught Me about Managing Risk," *Harvard Business Review*, Vol. 72, No. 2, pp. 76–90.

Shaw, Thomas E. (1990), "An Overview of Risk Management Techniques, Methods and Application," AIAA Space Programs and Technology Conference, September 25–27.

Smith, A. (1992), "The Risk Reduction Plan: A Positive Approach to Risk Management," IEEE Colloquium on Risk Analysis Methods and Tools.

Sprent, Peter. (1988), *Taking Risks: The Science of Uncertainty*, Penguin, New York.

Stone, J.R. et al. (1991), "Managing Risk in Civil Engineering by Machine Learning from Failures," IEEE First International Symposium on Uncertainty Modeling and Analysis, IEEE Computer Society Press, Los Alamitos, CA, pp. 255–259.

Toft, Brian. (1994), *Learning from Disasters*, Butterworth-Heinemann, Oxford, UK.

Wang, J. X. (1991), "Fault Tree Diagnosis Based on Shannon Entropy," *Reliability Engineering and System Safety*, Vol. 34, pp. 143–167.

Wang, J. X. (1996), "Complexity as a Measure of the Difficulty of System Diagnosis," *International Journal of General Systems*, Vol. 24, No. 3, pp. 257–269.

Wang, J. X. (2002), *What Every Engineer Should Know About Decision Making under Uncertainty*, CRC Press, Boca Raton, FL.

Wang, J. X. (2017), *Industrial Design Engineering: Inventive Problem Solving*, CRC Press, Boca Raton, FL.

Wang, J. X. (2019), "Complexity as a Measure of the Difficulty of System Diagnosis in Next Generation Aircraft Health Monitoring System," SAE Technical Paper 2019-01-1357, doi:10.4271/2019-01-1357.

Wang, J. X. (2019), "A Dynamic Fault Tree Approach for Time-Dependent Logical Modeling of Autonomous Flight Systems," SAE Technical Paper 2019-01-1358, doi:10.4271/2019-01-1358.

Wideman, R. Max., ed. (1992), *Project and Program Risk Management: A Guide to Managing Project Risks and Opportunities*, Project Management Institute, Drexel Hill, PA.

7 Cost Risk

Interacting with Engineering Economy

7.1 ENGINEERING: THE ART OF DOING WELL INEXPENSIVELY

Economics influences engineering design a great deal. When invoking the design process, the potential for over-budget cost is a major concern for every engineer. This chapter will look at several aspects of cost risk and how each one is associated with design.

Contrary to public belief, engineering embraces more than technology, automation, robots, and computers. In fact, cost risk assessment and management is part of all engineering design decisions. A railroad engineer in the nineteenth century defined engineering generally as "the art of doing well for one dollar, what any bungler can do with two". In order for their design to be profitable, engineers take hundreds of specifications into consideration. The specifications chosen are those which are required to satisfy the demands of the customer (i.e., general public or a plant requirement) and thus yield the maximum profit.

Money is the backbone of all business transactions and decisions, where engineering economy plays an important part. Money is the incentive to produce, improve, and compete, and it is also the means to do so. Analyzing the input and generation of money in the production process allows manufacturers to determine the success and effectiveness of their design policies, products, and ultimately their corporation at any point in time. Companies can use these analytical reports to guide future decisions.

Like food, warmth, and shelter, people have always needed water. A considerable body of specialized engineering knowledge and techniques has developed over the centuries to provide, control, treat, and dispose of water for ever-growing communities of people. The layout of a water supply system has no unique solution (Figure 7.1). The start and finish of the route are generally given by the location of the places between which water is to be moved, and it is the job of the engineer to select an economical, effective, and acceptable route from among the infinite possibilities between the termini.

Consider the problem of getting water from a source of supply on one side of a mountain, where the rain falls heavily, to the reservoir serving a city on the leeward side, which receives little rainfall. There are numerous ways that the water can be transported from the source, located at A, to the reservoir, located at F.

For example, a pump might be used to raise the water to a certain level and a tunnel could carry it through the mountain. The tunnel should exit the mountain no lower than the maximum reservoir level desired, which is determined by the

DOI: 10.1201/9781003371014-7

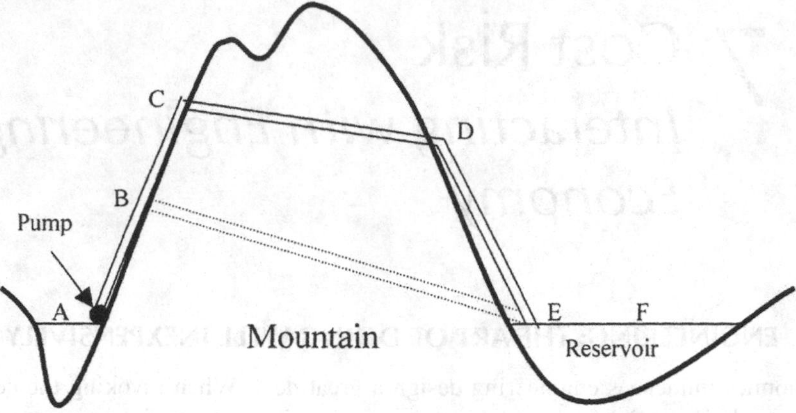

FIGURE 7.1 The elements in a pump/tunnel design for a water supply system.

geometry of the reservoir and the volume of water that it is designed to hold. The cross-sectional shape and size of the tunnel are related to the desired flow rate and can be analyzed with tools of the engineering science of hydraulics. If any kind of solids will be suspended in the water passing through the tunnel, its slope should be sufficiently steep to maintain a velocity that will prevent precipitation of the solids and the deposition of silt, which might in time clog the tunnel.

The total length of the tunnel will depend not only on its slope but also on exactly where it cuts through the mountain, which will have different kinds of rock at different locations. The different geological conditions will affect how fast the tunneling process can proceed, which, in turn, will affect the cost. All such construction costs are summarized in Figure 7.2 by the curve AB, which is exponential rather than linear because of the complex way in which the tunnel entrance height, H, contributes to the tunnel cost:

$$\text{Tunnel Cost} = 75000 \cdot \text{EXP}(-H/5)$$

The slope of the tunnel affects where its entrance is located, with steeper slopes requiring that the source water be pumped to higher elevations before gravity will carry it through the tunnel. Larger capacity pumps naturally cost more to purchase, install, and operate. The pumping cost over the mission time for this example is shown as the curve CD in Figure 7.2:

$$\text{Pumping Cost} = 40000 + 3000 \cdot H$$

By adding the ordinates of the curves representing the cost of tunneling and the cost of pumping, the combined-cost curve EG can be constructed:

$$C_T = \text{Pumping Cost} + \text{Tunnel Cost}$$

$$C_T = 40000 + 3000 \cdot H + 75000 \cdot \text{EXP}(-H/5)$$

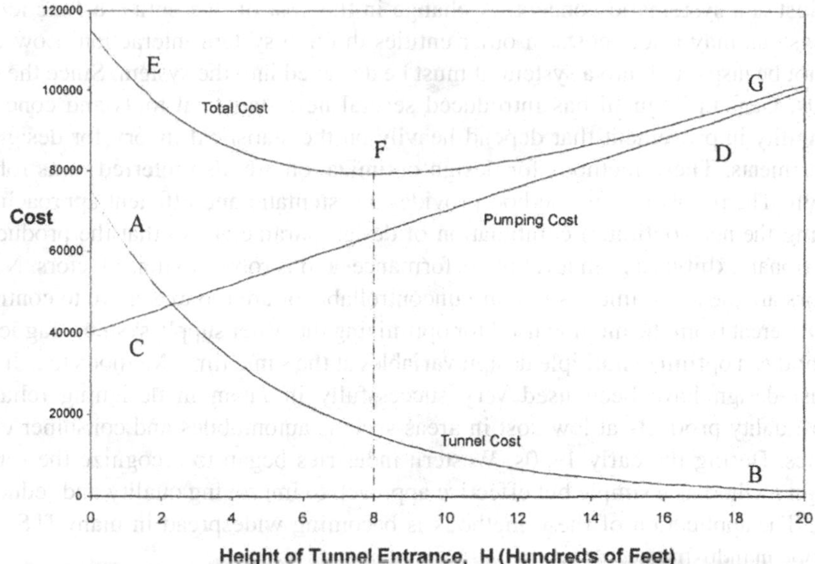

FIGURE 7.2 Cost curve for designing the water supply system (in thousands of dollars).

The optimal pump/tunnel combination, as determined by the criterion of minimal cost, is shown by point *F* on the combined-cost curve. Clearly, *H* is the only decision variable in this simplified example. Differentiating C_T with respect to *H*, we obtain the optimal condition (zero slope) as follows (Laptels rule):

$$\frac{dC_T}{dH} = 3000 + 15{,}000exp\left(\frac{-H}{5}\right) = 0$$

Thus,

$$H = 5 \cdot ln(5) = 8.05 \times 10^2 \text{ feet}$$

The example above is a simple case involving only one variable, the slope, for design optimization. Cost is the measure in monetary units of the work associated with the human endeavor. Current resource expenditure level is the measure in monetary units of the cost associated with human endeavor. All real systems have an allocated cost, which is a measure of the human endeavor required to conceptualize, evaluate, design, prototype, test, produce, deploy, operate, support, evolve, retire, and manage the system. Value is a monetary reward that drives consumer decisions and value is the measure of consumer choice. For an engineering project, the primary coordinates of value will include technical merit, cost, and market timing.

We have reduced all measures of work and value to cost. From the perspective of an individual, cost is a measure of value in that a consumer is willing to trade money for value. From the perspective of a project, a project is driven by cost. As time proceeds, the cost of a project grows and the residual value to be realized decreases.

Cost is a systematic concept. A change in the cost of one entity, a function or subsystem, may affect costs in other entities through system interaction. Low cost cannot be inspected into a system; it must be designed into the system. Since the late 1950s, Genichi Taguchi has introduced several new statistical tools and concepts of quality improvement that depend heavily on the statistical theory for design of experiments. These methods for design optimization are also referred to as robust design. The robust design method provides a systematic and efficient approach for finding the near-optimum combination of design parameters so that the product is functional, exhibits a high level of performance, and is robust to noise factors. Noise factors are those parameters that are uncontrollable or are too expensive to control.

Different from the method used for optimizing the water supply system, Taguchi's method can optimize multiple design variables at the same time. Methods to achieve robust design have been used very successfully in Japan in designing reliable, high-quality products at low cost in areas such as automobiles and consumer electronics. During the early 1980s, Western industries began to recognize the robust design method as a simple but effective approach to improving quality and reducing cost. The application of these methods is becoming widespread in many U.S. and European industries.

This chapter will briefly describe the robust design method for optimizing product design for cost and illustrate its application by the use of a design example.

7.2 TAGUCHI'S ROBUST DESIGN: MINIMIZE TOTAL COST

Delivering reliable high-quality products and processes at low cost has become the key to survival in today's global economy. Driven by the need to compete on cost and performance, many quality-conscious organizations are increasingly focusing on the optimization of product design. This reflects the realization that quality cannot be achieved economically through inspection. Designing in quality is cheaper than trying to inspect and re-engineer it after a product hits the production floor or worse, after it gets to the customer. Thus, new philosophy, technology, and advanced statistical tools must be employed to design high-quality products at low cost.

There is no single cost for a project until after its completion. In all cases prior to completion, a range and distribution of final cost exists which corresponds to the many possible outcomes resulting from future decisions and future uncertain outcomes. The premise for cost-risk analysis is that the distribution of possible costs is defined by future project decisions, although these decisions are not yet known.

The early design phase of a product or process has the greatest impact on life-cycle cost (LCC) and quality. Therefore, significant cost savings and improvements in quality can be enabled by optimizing product designs. The three major steps in designing a quality product are system design, parameter design, and tolerance design.

System design is the process of applying scientific and engineering knowledge to produce a basic functional prototype design. The prototype model defines the configuration and attributes of the product undergoing analysis or development. The initial design may be functional but it may be far from optimum in terms of quality and cost.

The next step, parameter design, is an investigation conducted to identify the settings of design parameters that optimize the performance characteristic and reduce the sensitivity of engineering designs to the sources of variation (noise). Parameter design requires some form of experimentation for the evaluation of the effect of noise factors on the performance characteristic of the product defined by a given set of values for the design parameters. This experimentation aims to select the optimum levels for the controllable design parameters such that the system is functional, exhibits a high level of performance under a wide range of conditions, and is robust to noise factors.

Experimenting with the design variables one at a time or by trial and error until a first feasible design is found is a common approach to design optimization. However, this approach is inefficient and can lead to either a very long and expensive time span for completing the design or a premature termination of the design process due to budget and schedule pressures. The result in most cases is a product design which may be far from optimal. As an example, if the designer is studying 13 design parameters at three levels, varying one factor at a time would require studying 1,594,323 experimental configurations (3^{13}). This is a "full factorial" approach where all possible combinations of parameter values are tried. Obviously, the time and cost involved in conducting such a detailed study during advanced design is prohibitive.

In contrast, the robust design method provides the designer with a systematic and efficient approach for conducting experimentation to determine near-optimum settings of design parameters for performance and cost. The robust design method uses orthogonal arrays (OAs) to study the parameter space, usually containing a large number of decision variables, with a small number of experiments. Based on design of experiments theory, Taguchi's OAs provide a method for selecting an intelligent subset of the parameter space. Using OAs significantly reduces the number of experimental configurations. A typical tabulation of one standard OA is shown in Table 7.1.

In this array, the columns are mutually orthogonal. That is, for any pair of columns, all combinations of factor levels (1, 2, or 3) occur, and they occur an

TABLE 7.1
The L9 (3^4) Orthogonal Array

	A	B	C	D
1	1	1	1	1
2	1	2	2	2
3	1	3	3	3
4	2	1	2	3
5	2	2	3	1
6	2	3	1	2
7	3	1	3	2
8	3	2	1	3
9	3	3	2	1

TABLE 7.2
Common Orthogonal Arrays with a Number of Equivalent Full Factorial Experiments Given in the Right Column

Orthogonal Array	Factors and Levels	No. of Experiments
L4	3 Factors at 2 levels	8
L8	7 Factors at 2 levels	128
L9	4 Factors at 3 levels	81
L16	15 Factors at 2 levels	32,768
L27	13 Factors at 3 levels	1,594,323
L64	21 Factors at 4 levels	4.4×10^{12}

equal number of times. Here, there are four factors A, B, C, and D, each at three levels. This is called an "L9" design, where 9 indicates the nine rows, configurations or prototypes to be tested, with test characteristics defined by the row of the table, i.e., a single row indicates the levels to be used for each factor in one experiment.

The number of columns of an OA represents the maximum number of factors that can be studied using that array. Note that this design reduces 81 (3^4) configurations to 9. Some of the commonly used OAs are shown in Table 7.2. As Table 7.2 shows, there are greater savings in testing for the larger arrays. This reduction in testing comes with a lack of information about the interactions between the factors.

Using an L9 OA means that nine experiments are carried out in search of the combination of four factor levels which will result in the near-optimal mean, and also the near-minimum variation away from this mean. To achieve this, the robust design method uses a statistical measure of performance called signal-to-noise (S/N) ratio borrowed from electrical control theory. Taguchi's S/N ratio is a performance measure to choose control levels that best cope with noise. The S/N ratio takes both the mean and the variability into account. In its simplest form, the S/N ratio is the ratio of the mean (signal) to the standard deviation (noise). The S/N equation depends on the criterion for the quality characteristic to be optimized. While there are many different formats for S/N ratios, three of them are considered standard and are generally applicable in the following situations:

- Biggest-is-best quality characteristic (strength, yield);
- Smallest-is-best quality characteristic (contamination);
- Nominal-is-best quality characteristic (dimension).

Whatever the type of quality or cost characteristic, the transformations are such that the S/N ratio is always interpreted in the same way, where the larger the S/N ratio, the better.

By making use of OAs, the robust design approach improves the efficiency of generating the information necessary to design systems that are robust to variations in manufacturing processes and operating conditions. As a result, development time can be shortened and R&D costs can be reduced considerably. Furthermore, a near-optimum choice of parameters may result in wider tolerances so that lower cost components and production processes can be used.

The third step, tolerance design, is the process of determining tolerances around the nominal settings identified in the parameter design process. Tolerance design is required if robust design cannot produce the required performance without costly special components or high-accuracy processes. It involves tightening of tolerances on parameters where their variability could have a large negative effect on the final system. Typically, tightening tolerances leads to higher cost.

Traditionally, engineers focus on system and tolerance design to achieve performance. The common practice in product and process design is to base an initial prototype on the first feasible design (system design). Then, the reliability and stability against noise factors are studied and any problems are corrected by requesting costlier components with tighter tolerances (tolerance design). In other words, parameter design is largely overlooked. As a result, the opportunity to improve quality while decreasing cost is usually missed.

Recently, however, the use of Taguchi's robust design methods have been increasing within industry. Many companies are now realizing that new tools are required for survival in the increasingly competitive global marketplace. Thus, it is expected that the application of these methods will become widespread as low LCC, operability, and quality issues replace performance as the driving design criteria.

Optimizing a product or process design means determining the best system architecture, optimum settings of control factors, and tolerances. Robust design is Taguchi's approach for finding near-optimum settings of the control factors to make the product insensitive to noise factors. The five steps for the conduct of the robust design method are summarized as follows:

1. Identify the system function and noise factors,
2. Identify the total cost function and control factors,
3. Design the matrix of experiments and define the data analysis procedure,
4. Conduct the experiments and data analysis,
5. Predict the cost risk at these levels.

7.3 STEP 1: IDENTIFY SYSTEM FUNCTION AND NOISE FACTORS

The details of the five steps in robust design are described using a simple heat exchanger design-for-cost example. The approach is equally applicable to quality characteristics to be optimized other than cost, such as weight, yield, processing time, structural strength, or surface defects.

The system function of the compressed air cooling system shown in Figure 7.3 is to cool the air from 95°C to 10°C between the two stages of compression. The system first cools the air in a pre-cooler and then in a refrigeration unit. Water goes through the condenser of the refrigeration unit, then into the pre-cooler, and finally into a radiator where the heat is ejected.

The flow rate of compressed air is given as 1.2 kg/s and the flow rate of water is 2.3 kg/s. The water is expected to leave the radiator at a temperature of 24°C. The

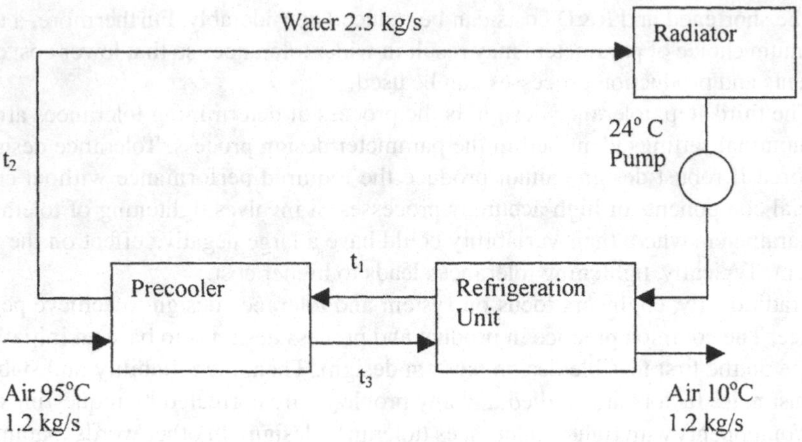

FIGURE 7.3 Compressed air cooling system example.

system is to be designed for minimum total cost, where this cost is the sum of the costs in dollars of the refrigeration unit, the pre-cooler, and the radiator. The parametric cost equations (X_i) for the refrigeration unit, the pre-cooler, and the radiator in terms of the output temperatures (t_i) are given below.

$X_1 = 1.20\, a\, (t_3 - 10)$,
$X_2 = 1.20\, b\, (95 - t_3)/(t_3 - t_1)$ for ($t_3 > t_1$),
$X_3 = 2.3\, c\, (t_2 - 24)$, where
X_1 = cost of refrigeration unit (\$), a = cost parameter for refrigerator,
X_2 = cost of pre-cooler (\$), b = cost parameter for pre-cooler,
X_3 = cost of radiator (\$), c = cost parameter for radiator.

Assume the values of the cost parameters are $a = 48$, $b = 50$, and $c = 25$. Now that the system function for the heat exchanger is identified, the next step in robust design is to identify the noise factors.

Noise factors are those that cannot be controlled or are too expensive to control. Examples of noise factors are variations in operating environments, variation in materials, and manufacturing imperfections. Noise factors cause variability and loss of quality. The overall aim is to design and produce a system that is insensitive to noise factors. The designer should identify as many noise factors as possible and then use engineering judgment to decide the more important ones to be considered in the analysis.

Various noise factors may exist in the air cooling process example under consideration. For this example, we will assume that engineers have identified three important noise factors:

N_1 = cost parameter for refrigeration unit,
N_2 = radiator output temperature,
N_3 = input temperature of compressed air.

The first noise factor N_1 is taken to be the cost parameter for refrigerator (a) since the refrigeration unit cost was uncertain at the time. It was decided that this factor could be higher than the original estimate of 48 (up to 56). The second noise factor N_2 is the radiator output temperature of 24°C. This temperature could vary depending on environmental factors and could be higher, up to 27°C. The last noise factor N_3 is the input temperature of the compressed air coming in at a temperature of 95°C. This temperature could also vary depending on the operating conditions, up to 100°C.

7.4 STEP 2: IDENTIFY TOTAL COST FUNCTION AND CONTROL FACTORS

In this example, designing for cost is the objective. Therefore, the objective function to be optimized is the total cost (TC) of the system:

Minimize $TC = X_1 + X_2 + X_3$
Subject to: compression, power, energy, and mass balance constraints.

The objective now is to find the design that minimizes total cost, considering the uncertainty due to the noise factors cited above.

In this example, output temperatures t_1, t_2, and t_3 are the control factors since they can be changed by the designer to determine the sizes, hence the cost of pre-cooler, refrigeration unit, and the radiator.

After a preliminary study, an initial set of values were determined for the control parameters under ideal conditions, i.e., without considering the noise factors

$t_1 = 28°C$,
$t_2 = 39°C$, and
$t_3 = 38°C$.

As a next step, the design engineers and cost analysts desire to study alternative levels for the controllable design parameters considering the uncertainty due to noise factors. In robust design, generally, two or three levels (or settings) are selected for each factor. The level of a test parameter refers to how many test values of the parameter are to be analyzed. Commonly, one of these levels is taken to be the initial operating condition. These levels should be taken sufficiently far apart so that the chance is increased for capturing any non-linearity of the relationship between the control factors and the noise factors.

For the example case, three alternative levels were identified to be studied for the controllable design factors as shown in Table 7.3(a). These factor levels define the experimental region to be studied. All of these levels should satisfy the power, energy, and mass balance constraints. Level two represents the initial setting (base line) for the control factors.

The three noise factors were chosen and studied at two levels. These levels were identified and they are given in Table 7.3(b), where level one represents the initial setting for the noise factors.

TABLE 7.3
Factor Levels for the Optimization Problem Control Factor Levels Noise Factor Levels

Control Factor Levels

	1	2	3
t_1	25	28	31
t_2	36	39	42
t_3	35	38	41

Initial Setting
(a)

Noise Factor Levels

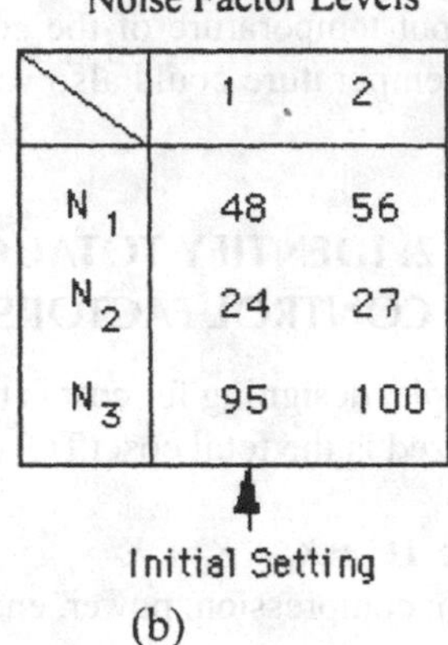

	1	2
N_1	48	56
N_2	24	27
N_3	95	100

Initial Setting
(b)

7.5 STEP 3: DESIGN MATRIX OF EXPERIMENTS AND DEFINE DATA ANALYSIS

The objective now is to determine the optimum levels of control factors so that the system is robust to noise factors. Robust design methodology uses OAs based on the design of experiments theory to study a large number of decision variables with a relatively small number of experiments. Using OAs significantly reduces the number of experimental configurations required.

7.5.1 Construct Orthogonal Arrays

To select the appropriate OA to fit a specific case study, we need to count the total degrees of freedom to find the minimum number of experiments that must be performed to reach a near-optimum parameter set. One degree of freedom is associated with the overall mean regardless of the number of control factors. To this, we add the degrees of freedom associated with each control factor, which is equal to one less than the number of levels. For our example, we have

Factor Degrees of Freedom

Overall mean = 1
t_1 3 − 1 = 2
t_2 3 − 1 = 2
t_3 3 − 1 = $\underline{2}$
Total = 7

Therefore, we need to conduct at least seven experiments to reach a near-optimum case in our example. This fits nicely into Taguchi's standard L9 array, shown in

TABLE 7.4
L4 (2^3) Orthogonal Array

	A	B	C
1	1	1	1
2	1	2	2
3	2	1	2
4	2	2	1

Table 7.1. In order for an array to be a viable choice, the number of rows must at least be equal to the degrees of freedom required for the case study. The L9 array has eight degrees of freedom and it can handle four factors at three levels. Since we have only three control factors, one of the columns of the array will be left empty. Orthogonality is not lost by keeping one or more columns of an array empty.

Using the same procedure, an L4 array (Table 7.4) was selected for the three noise factors from the tabulated standard OAs.

7.5.2 Orthogonal-Array-Based Simulation

After selecting the appropriate OAs, the next step is to develop a procedure to simulate the variation in the quality characteristic, total cost in our example, due to the noise factors. A common approach is the use of Monte Carlo simulation. However, for an accurate estimation of the mean and variance, Monte Carlo simulation requires a large number of experimental testing conditions which can be expensive and time-consuming. As an alternative, Taguchi proposes orthogonal-array-based simulation to evaluate the mean and the variance of a product's response resulting from variations in noise factors. With this approach, OAs are used to sample the domain of noise factors. The diversity of noise factors is studied by crossing the OA of control factors by an OA of noise factors. Thus in this case, we evaluate total cost (TC) for each of the nine trials against the background of four different combinations of noise conditions, as shown in Table 7.5. The results of the experiments are denoted by $Y_{i,j}$.

Thus, using an orthogonal-array-based simulation algorithm as shown in Table 7.5, one can study four control factors against the background of three noise factors, by running only 36 experiments as opposed to 324 ($3^4 \times 4$) required by a full factorial approach.

7.6 STEP 4: CONDUCT EXPERIMENTS AND DATA ANALYSIS

The robust design method can be used in any situation where there is a controllable process. The controllable process can be an actual hardware experiment. Conducting an actual hardware experiment can be costly. However, in most cases, systems of mathematical equations can adequately model the response of many products and processes. In such cases, these equations can be used adequately to conduct the controlled

TABLE 7.5
Orthogonal-Array-Based Simulation Algorithm

Noise Orthogonal Array

	1	2	3	4
N_1	1	1	2	2
N_2	1	2	1	2
N_3	1	2	2	1

Control Orthogonal Array

	A	B	C	D	$Y_{i,j}$			
1	1	1	1	1				
2	1	2	2	2				
3	1	3	3	3				
4	2	1	2	3				
5	2	2	3	1				
6	2	3	1	2				
7	3	1	3	2				
8	3	2	1	3				
9	3	3	2	1				

matrix of experiments. Sometimes, these equations have a simple closed solution, as in our case, but in most cases, computer models must be used to solve them.

For our example, the matrix experiment given in Table 7.5 is conducted using the appropriate system of mathematical equations for cost, compression power, energy, and mass balance constraints. The response ($Y_{i,j}$), which is the total cost in dollars for this case, is computed for each combination of control and noise matrix experiments. The results are displayed in Table 7.6. Note that column C in the control OA was left empty since we have only three variables in this example. For each combination of control factor levels, the mean cost (Mean) and the standard deviation (Std) are also shown in Table 7.6.

The traditional analysis performed with data from a designed experiment is the analysis of the mean response. The robust design method employs an S/N ratio to include the variation of the response.

7.6.1 Signal-to-Noise Ratio

The S/N developed by Dr. Taguchi is a statistical performance measure used to choose control levels that best cope with noise. The S/N ratio takes both the mean and the variability into account. The particular S/N equation depends on the criterion for the quality characteristic to be optimized. Whatever the type of quality

TABLE 7.6
Results of the Matrix of Experiments

Noise Matrix

	1	2	3	4
N_1	48	48	56	56
N_2	24	27	24	27
N_3	95	100	100	95

Control Matrix

	t_1	t_2	C	t_3	$Y_{i,j}$				Mean	Std
1	25	36		35	4691	3998	4961	4208	4465	441
2	25	39		38	5489	4790	5782	5036	5274	445
3	25	42		41	6325	5621	6641	5899	6122	451
4	28	36		41	4926	4226	5247	4501	4725	452
5	28	39		35	5568	4888	5851	5086	5348	440
6	28	42		38	6291	5598	6590	5838	6079	446
7	31	36		38	4993	4312	5304	4539	4787	447
8	31	39		41	5723	5031	6051	5298	5526	452
9	31	42		35	6677	6029	6991	6194	6473	442

characteristic, the transformations are such that the S/N ratio is always interpreted in the same way with the larger the S/N ratio the better.

The smallest-is-best-type S/N ratio will be used to analyze the results, since our objective here is to minimize cost. The smallest-is-best S/N ratio is defined as S/N = –10 *log(Σ(Y^2)/n)) as explained below.

Since log is a monotone function, maximizing S/N is equivalent to minimizing the quality characteristic. The S/N ratios for the data of our example are computed and displayed in Table 7.7(a). As an example, S/N for experiment 1, for the four observations under different noise conditions (n=4), is computed as follows:

$$\text{S/N} = -10 \log \left\{1/4\left[(4691)^2 + (3998)^2 + (4961)^2 + (4208)^2\right]\right\} = -73.03$$

7.6.2 Data Analysis Using the S/N

There are several approaches to this analysis. One common approach is to use statistical analysis of variance (ANOVA) to see which factors are statistically significant. Dr. Taguchi suggests another approach, which involves graphing the effects and visually identifying the factors that appear to be significant. We will follow this simpler approach for the analysis.

TABLE 7.7
Signal-to-Noise Ratios and the Response Table

Control Matrix

	t_1	t_2	C	t_3	S/N
1	1	1	1	1	-73.03
2	1	2	2	2	-74.47
3	1	3	3	3	-75.76
4	2	1	2	3	-73.52
5	2	2	3	1	-74.59
6	2	3	1	2	-75.70
7	3	1	3	2	-73.63
8	3	2	1	3	-74.87
9	3	3	2	1	-76.24

(a) Signal to Noise Ratio

	Average S/N		
	t_1	t_2	t_3
1	-74.42	-73.39	-74.62
2	-74.60	-74.64	-74.60
3	-74.91	-75.90	-74.72

(b) Response Table

Since the experimental design is orthogonal, it is possible to separate out the effect of each factor. The average S/N ratios for each level of the three control factors are displayed in the response table given in Table 7.7(b). The S/N ratios shown in the response table are calculated by taking the average from Table 7.7(a) for a parameter at a given level every time it was used. As an example, t_2 was at level two in experiments 2, 5, and 8. The average of corresponding S/N ratios is 74.64 which is shown in the response table under t_2 at level 2.

The average S/N ratios from the response table are plotted in Figure 7.4. The graphs reveal that control parameter t_2 is more significant than t_1 and t_3, having the largest effect on cost. Clearly, level one appears to be the best choice for parameters t_1 and t_2 since it corresponds to the largest average S/N ratio. As for t_3, there is not much difference between levels one and two; however, level two seems to be the best choice.

As a result of the above analysis, the near-optimum levels for the three controllable parameters were selected as follows:

Test Parameters: t_1 t_2 t_3
Optimum Levels: 1 1 2
Parameter Setting (C): 25 36 38

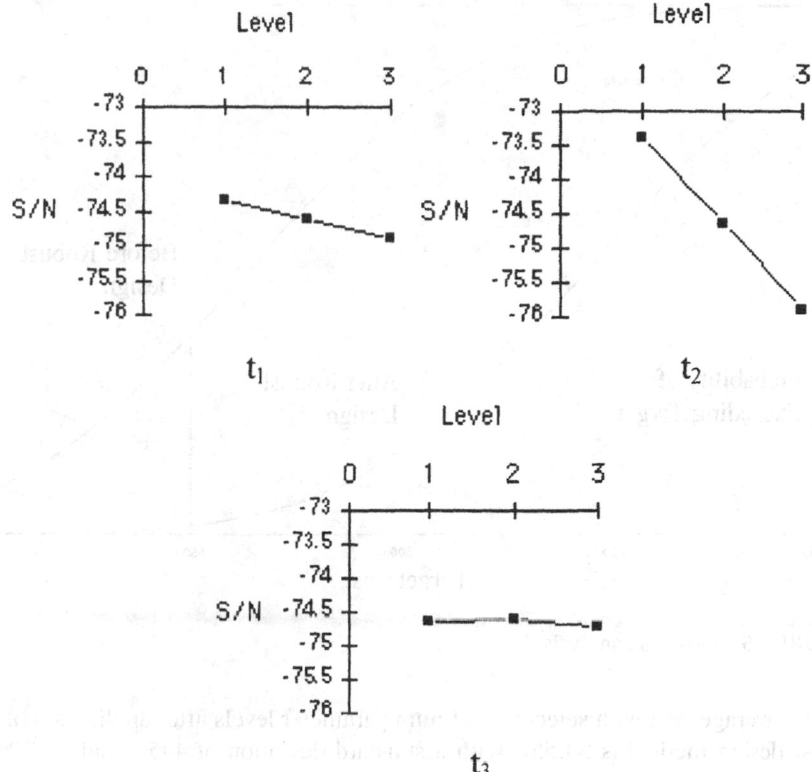

FIGURE 7.4 Graphs of S/N ratios.

Note that this combination was not considered during the nine experiments carried out. In robust design, the predicted near-optimum setting need not correspond to one of the rows of the matrix experiment.

7.7 STEP 5: PREDICTION OF COST RISK UNDER SELECTED PARAMETER LEVELS

Recall that the initial (baseline) settings for the control parameters before the optimization process were as follows:

Test Parameters: t_1 t_2 t_3
Parameter Level: 2 2 2
Parameter Setting (C): 28 39 38

These settings correspond to an initial average cost of $5,357 with a standard deviation of 445.6 and an S/N of –74.60. Assuming a normal distribution and a target cost of $5,500, the probability of exceeding the target value is 37.4%. A significant cost risk exists (see Figure 7.5).

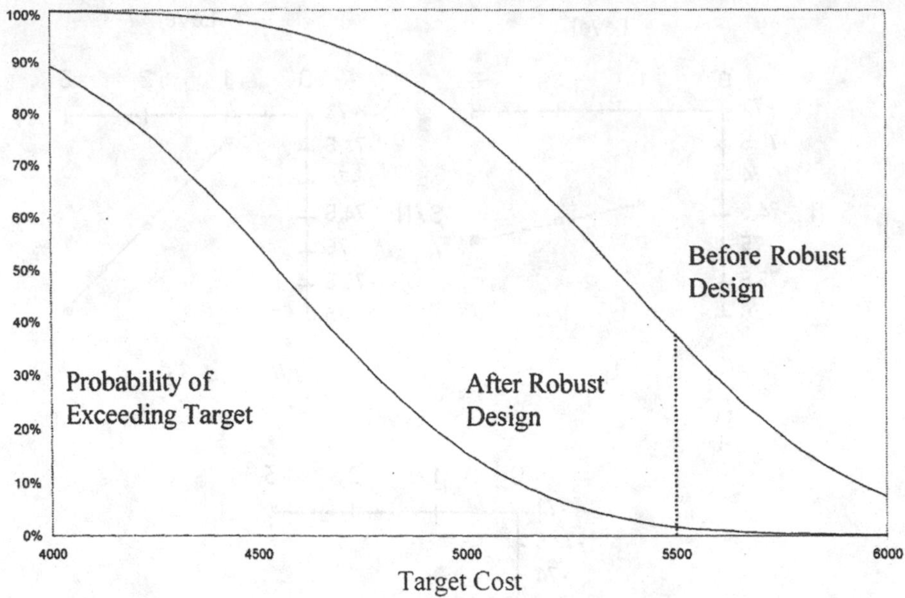

FIGURE 7.5 Cost-risk analysis.

The average cost with selected optimum parameter levels after application of the robust design method is $4,551, with a standard deviation of 445.4 and an S/N of –73.19. Therefore, in this case, robust design resulted in a cost saving of $806 (about 15%) and a slight reduction in standard deviation. Again assuming a normal distribution and a target cost of $5,500, the probability of exceeding the target cost is reduced to 1.7%.

The robust design method is a systematic and efficient approach for determining the near-optimum configuration of design parameters for cost. Principal benefits include considerable time and resource savings; determination of important factors affecting operation, performance, and cost; quantitative measures of the robustness (sensitivity) of the near-optimum results; and quantitative recommendations for design parameters that achieve the lowest-cost, high-quality solutions.

Robust design moves cost considerations to the design stage where the greatest benefits can be derived. Also, the robust design approach can aid in integrating cost and engineering functions. Overall, results suggest that robust design is a powerful tool which offers simultaneous improvements in quality, cost, and engineering productivity.

7.8 LIFE-CYCLE COST MANAGEMENT (LCCM)

The life-cycle of a system/product is made up of the following major phases:

Specification;
Design;

Production/Construction;
Use and Retirement.

LCCs are product (or system or program) costs from inception to disposal. The objective of managing LCC is to choose the most cost-effective approach from a series of alternatives so that the lowest long-term cost of ownership is achieved. The definitions for LCC and its assessment include:

LCCs – The total costs of a project over it's lifetime. These total costs are the sum of the initial project costs (research, development, design, construction, startup, etc.) plus the operating costs (maintenance and operations) over the expected life of the project.

LCC Assessment – A general approach to economic evaluation of LCC, which takes into account all dollar costs related to owning, operating, maintaining, and disposing of a product/project over the appropriate study period.

Today, government procurements are driven by Best Value and Total Ownership Costs, and LCCM is recognized as a powerful tool for both contractor and customer in achieving total product-cost objectives. According to Kohoutek (1996), "the Life-Cycle Cost of a system is the total cost to the government of acquisition and ownership of that equipment over its entire life".

The first section of ISO 14040, a part of the international environment protection standard (ISO 14000), deals with the principles and application of Life-Cycle Assessment. The American National Standards Institute; CEN in Europe; and NORSOK, a standard for the Norwegian offshore oil industry, each uses LCC as the basis for making investment decisions on plant and equipment.

7.8.1 Why Life-Cycle Cost Management?

One of the primary benefits of past applications of LCC methodology is the considerable evidence showing that by the time a product is ready to be released to manufacturing, a substantial portion of its eventual LCC is already locked in, even though it has not been accrued yet.

Like the NORSOK model, most LCC programs have been designed for specific industries and processes. For example, the LCC models for pumping systems can be found in ANSI/SAE ARP 4293 standard. Published in 1992, this standard was developed originally for military aircraft and is now gaining recognition as an American national standard.

An important fundamental principle of LCC is the understanding that the lowest manufacturing cost does not necessarily yield the lowest LCC to either the manufacturer or the customer. Let's take a pump as an example. In the world of hard-to-pump fluids, for example, the purchase cost can fade to insignificance when compared with the operating costs over the life of the pump. In these difficult applications, the cost of excessive wear, maintenance, spare parts, unplanned downtime, loss of productivity, seal replacements, and product damaged by the pump forms a substantial proportion of the LCC, dwarfing the capital expenditure and routine operating costs.

To purchase equipment like a pump for hard-to-pump fluids, a customer has to find a balance between the contemplated mission requirements and budget resources, taking into account the internal constraints given by

Internal acquisition logistics: for example, status of incoming inspection and testing facilities, installation opportunity, asset management system and procedures.

The support system, which will affect the actual application environment stresses, maintenance strategy, etc.

The supplier, who wants to satisfy the proposed mission requirements, must perform many analyses to gain insight into available implementation alternatives and find the one that guarantees the achievement of these requirements at minimum cost. Major constraints for this activity depend on available

Cost-estimating skills and completeness of the historical cost database;
Tools, skills, information, and resources for accurate analysis;
Creative design and implementation alternatives.

In summary, the benefits of carrying out an LCC analysis are as follows:

To provide justification for "spend to save" decisions;
Enable comparison of competing systems;
Allow evaluation of alternative systems like pumping rather than conveying;
Enable better decisions;
Enable monitoring a program or process more effectively; and
Allow measuring the impact of different levels of reliability and maintainability to facilitate trade-off decisions with other priorities.

7.8.2 Life-Cycle Cost Analysis Theory Is Simple

It is the sum of all monies spent, received, and attributed directly and indirectly to a defined system from its inception to its dissolution. This definition encompasses the acquisition, ownership, and disposal phases. Turning now to the phases of LCC for comparing two or more pump systems, the first stage is to identify the key cost drivers for the process being evaluated (Table 7.8). Does the fluid to be pumped have one or more of the characteristics listed in the table? How critical is downtime and what is its effect on productivity?

Now, consider the effect of the key cost drivers on the existing pump system and estimate the costs associated with each factor. How much do you spend for spare parts in a month, or in a year? What is the labor cost for dismantling, repairing, unclogging, and carrying out unplanned maintenance? What is the cost of downtime in terms of deferred production? How many seals does pump pulsation and high radial and axial loads destroy in a year?

In the case of delicate and shear-sensitive products, there is also the cost of product degradation to consider. Although this can be the largest, most significant factor

TABLE 7.8
Life-Cycle Cost Drivers for Hard-to-Pump Fluids Pump

Hard-to-Pump Fluids	Cost Driver
Highly abrasive fluids	Premature wear of pumping mechanism with "impingement" devices, centrifugal and PC pumps.
Slurries with solid debris	Premature wear, depending on nature of solid.
	Loss of capacity in centrifugal pumps, leading to downtime.
	Possible clogging in PC pumps, leading to high maintenance and downtime.
Fluids with viscosity greater than specification	Loss of capacity in centrifugal pumps, leading to downtime.
Fluids without enough air/gas	Excessive cavitation and vapor-locking, leading to high maintenance and downtime.
	Possibility of catastrophic failures in PC pumps.
Fluids with large and/or stringy solids	Clogging of mechanism, leading to high maintenance and downtime, with centrifugal and PC pumps.
Common pump problem	Cost Driver.
Pump pulsation	Shorten seal life.
	Excessive vibration to surrounding environments.
High radial and axial loads	Shorten seal life.
	Shaft fatigue; bearing fatigue.
Not able to run dry	Limits production flexibility.

in LCC, it is one that is nevertheless frequently overlooked. So many pump systems rely on an impingement principle – high amount of contact between the fluid and pump mechanism – that the degradation problem is seen as inevitable. But consider the handling of a crystal slurry. Pump-induced impingement damages as much as 40% of the product and changes the particle size distribution of the final product. This diminishes the retail value of the crystals and requires extra production to make up for the loss.

Placing an accurate value on each cost driver is another key to successful LCC analysis. Cost estimates are a mix of analysis of existing data and prediction of future costs. Estimating procedures first use statistical analysis of historical data and then use predictions based on experience gained from previous systems. As one becomes more familiar with LCC techniques and starts to accumulate a database of operational data, it becomes feasible to simulate the logistic elements. Simulation gives you the ability to investigate the interplay between random events and those with time-related variations to more explicitly evaluate uncertainty.

The estimates should be as accurate as possible, given the stage at which the assessment is being undertaken. One must also recognize and define risks and uncertainties, if possible. The analyst should seek to make the estimate consistent and appropriate. It is not necessary to apply a fully rigorous LCC analysis to arrive at a meaningful result.

Example 7.1

For a pumping system in a space station, a large safety index can reduce the probability of failure, but will increase the initial cost due to the weight penalty. The expected LCC can be calculated as:

$$E(C) = C_i + P_f C_f$$

where, C_i, the initial cost, can be calculated using the system's safety index, β,

$$C_i = 20\text{EXP}(0.2236\beta)$$

The probability of failure, P_f, can also be calculated using the safety index,

$$P_f = \phi(-\beta)$$

Suppose the cost of failure of the system is C_f = $10 million. Select the safety index, β, to minimize the expected LCC.

As shown in Figure 7.6, the design that minimizes expected LCC is

β = 3.47
C_i = $43,451
P_f = 0.00026
$E(C)$ = $46,053

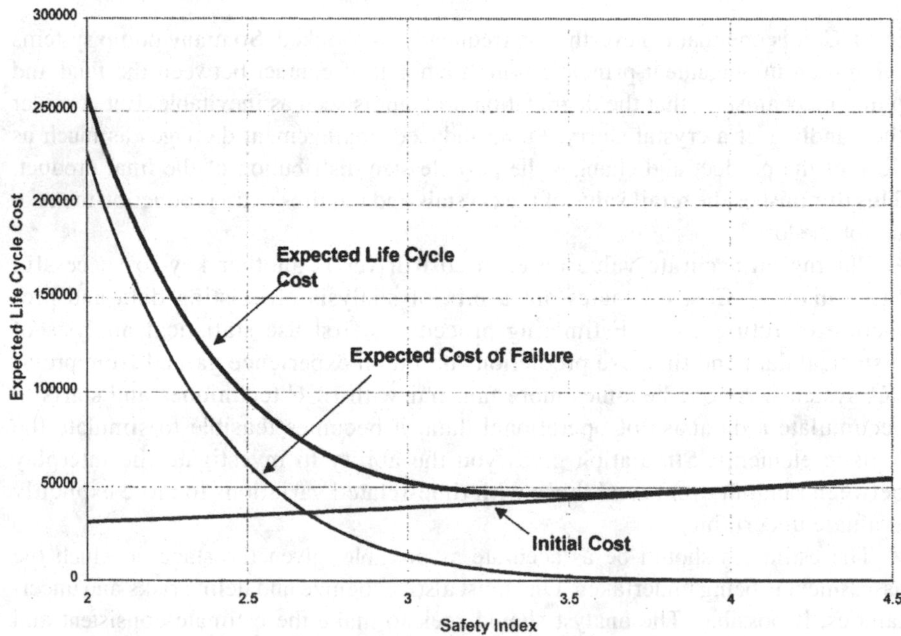

FIGURE 7.6 Life-cycle cost risk analysis for a pumping system in a space station.

7.8.3 Control Life-Cycle Cost Management Process

Use of the LCC is an idea whose time has come. Although LCC analysis was first proposed more than 25 years ago, it remained a theoretical concept mentioned in economics course, discussed at the academic level, but rarely applied in practice.

The two primary reasons for the slow development of a generic, defined process for LCC analysis and its incorporation into the broad practice of managerial decision making are:

- the perceived complexity of financial analysis techniques and/or processes associated with life-cycle analysis, and;
- the difficulty in determining and acquiring the appropriate data to be used in the analysis.

Obviously, one reason for not implementing an LCC program is that it is perceived as excessively time-consuming and complex. This is not necessarily an accurate assessment. If you limit LCC estimates to those few elements known to account for most of the total costs, the issue is easier. For hard-to-pump fluid applications, this includes the initial purchase cost, spare parts costs, labor for repairs and maintenance, costs associated with downtime and lost productivity, damaged products, seal life, and so on.

LCC analysis can be easy or difficult. The time spent collecting and analyzing the data for comparison purposes should be appropriate to the level of investment and project at hand. For a single pump purchase with no special metallurgies or design consideration, no more than an hour of preparation may be required. Of course, you should expect to make a more thorough analysis for multiple pump purchases or long-term strategic planning that requires a redesign of the existing production process.

Nevertheless, LCC does have a role in the smaller investment decisions. David Gess, an LCC consultant for the utilities industries, remarks, "Although major investment decisions often reflect total life-cycle costs, most firms make thousands of smaller decisions without life-cycle costs in mind… these costs can add up to a large part of operative expenses".

Although LCC is widely known and understood as a concept, there are several obstacles to implementing LCC programs in practice. One of the most significant barriers is the division between the capital budget and the maintenance and operating budget. The justification for handling capital and maintenance budgets separately is that it minimizes investment to boost short-term profit. However, the shortsightedness of this approach is becoming apparent as companies suffer the consequences of poor purchasing decisions. Later, they are increasingly required to justify high levels of maintenance spending.

Another barrier to implementing LCC used to be that it was time-consuming to gather the data required. Computers made this objection disappear. The latest analysis software makes LCC calculations easier and faster than ever before. Moreover, experience shows that companies using LCC regularly in their equipment purchases find substantial savings in time, as well as money, over the long term.

Despite the hurdles, LCC analysis is the only way to make informed decisions about maximizing productivity and profitability. The best advice on implementing an LCC program is simply not to make it more difficult than it has to be. It is about reducing the risk in investment decisions and justifying the use of common sense and experience. Remember that it is essentially about comparing two or more options, not about finding an absolute value for total costs. With this in mind, some options immediately drop out of the equation for your particular application.

The days of basing investment decisions solely on capital cost alone, while the maintenance budget remains an open checkbook, are coming to an end. Management looks increasingly at the total cost of ownership – another way of saying LCC – and recognizes the need to get the most out of equipment purchases. It is no longer a question of whether you can afford to carry out LCC analysis, but whether you can afford not to.

In today's competitive work, environment cost control is essential. Management discussions about planning and/or cost analysis activities within the organization increasingly include references to LCC analysis. The term is becoming more popular as more people are using it. It is interesting to note that there is rarely any mention of a "generic process" that one might associate with the term. A major reason for this is the non-uniformity of current-day LCC analysis practices. Each entity does it uniquely, with little or no agreement as to the most effective method for bringing relevant and vital information to the management decision making process.

Responsible engineering will minimize cost risk and the waste of scarce resources. Economic constraints are imposed by the competition of the marketplace. Engineers often experiment with the possibilities of improving their products while lowering the cost of manufacturing. A common device for this purpose is CAD (computer-aided design), which allows engineers to effectively synthesize a product in the computer while avoiding the cost and time of building a prototype. It is even possible to create a life-sized model using lasers and special computerized equipment to create plastic models for modestly sized items. Computerized simulation, which will be further explored in Chapter 9, reduces modeling time from weeks to hours.

7.9 PROBABILISTIC COST DRIVERS: QUANTIFYING COMPLEXITY OF PROJECT BUDGETING

This section will address risk-based probabilistic cost-estimating methods that can improve our appreciation of the cost of uncertainty and potential risk events. The identification and characterization of potential risks (threats or opportunities) that can affect outcomes as well as the inherent uncertainty in estimating the value of any future project element or process will be covered. For example, using a probabilistic approach, it is possible to better recognize this potential outcome in the bid phase to:

- Change the budget (as the owner) or proposal-bid value (as the contractor) if doing so is consistent with a strategic plan to win project funding (as the owner) or the bid (as the contractor), in order to complete the project on time and within budget (as the owner) or at a profit (as the contractor), or
- To stop the project (as an owner) or withdraw from the project (as a contractor) if a strategy to win the bid and still realize a profit is not feasible.

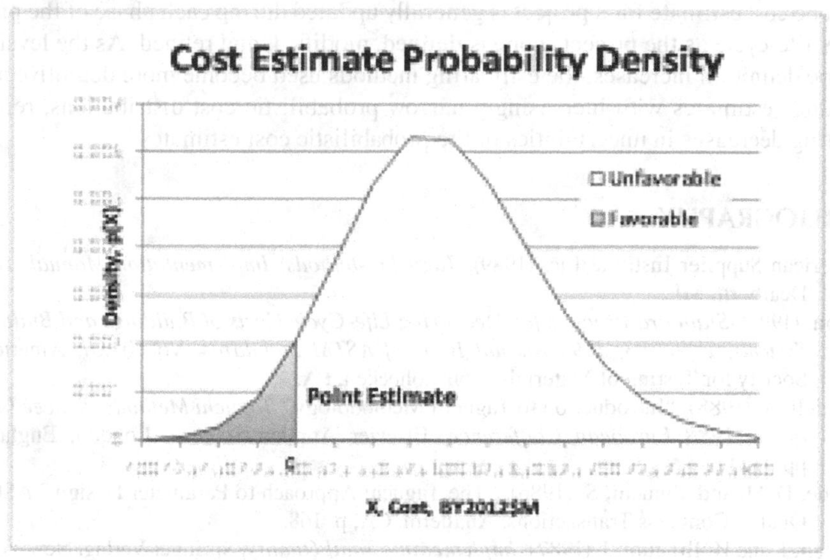

FIGURE 7.7 Probabilistic cost drivers: Quantifying complexity of project budgeting.

7.9.1 Common Probabilistic Cost Risk Modeling Methods

As illustrated in Figure 7.7, common probabilistic cost risk modeling methods can be grouped loosely into four types:

- Line item ranging, which applies a distribution to each line in an estimate or summary of an estimate, usually combined with discrete risk events which are described in terms of the
- likelihood of them arising and the effect they will have on the cost;
- Risk factor/risk driver models with a many-to-many relationship between risks and costs, where uncertainties are defined in terms of drivers that may individually affect several cost elements and where several drivers may act jointly on a single cost; and
- Approaches that are parametric or hybrid, such as a model that considers both the expected value of project-specific risks and the impact of systemic risk.

7.10 SUMMARY

In order to provide information that is pertinent to risk identification, characterization, and management, risk-based, probabilistic cost-estimating methodologies can be built upon a deterministic cost base and added consideration of variability and prospective risk occurrences. Additionally, risk-based, probabilistic cost-estimating methods provide more information for managing budgets (for owners) and securing a project in a setting of competitive bidding (for contractors), as well as inform strategies to manage disputes and claims in construction (owners and contractors).

The cost estimate for a project is generally updated during each phase of the project's life-cycle as the project scope is defined, modified, and refined. As the level of scope definition increases, the estimating methods used become more definitive and produce estimates with increasingly narrow probabilistic cost distributions, representing decreases in uncertainties of the probabilistic cost estimates.

BIBLIOGRAPHY

American Supplier Institute Inc. (1989), *Taguchi Methods: Implementation Manual*, ASI, Dearborn, ML.

Anon. (1994), *Standard Practice for Measuring Life-Cycle Costs of Buildings and Building Systems, E 917-93, 1994 Annual Book of ASTM Standards*, Vol. 04.07, American Society for Testing of Materials, Conshohocken, PA.

Bendell, A. (1988), "Introduction to Taguchi Methodology," *Taguchi Methods: Proceedings of the 1988 European Conference*, Elsevier Applied Science, London, England, pp. 1–14.

Bryne, D. M. and Taguchi, S. (1986), "The Taguchi Approach to Parameter Design," ASQC Quality Congress Transactions, Anaheim, CA, p. 168.

Cullen, J. and Hollingum, J. (1987), *Implementing Total Quality*, Springer-Verlag, New York, NY.

Fabrycky, W. J. and Blanchard, B. S. (1991), *Life-Cycle Cost and Economic Analysis*, Prentice Hall, Englewood Cliffs, NJ.

Gunter, B. (1987), "A Perspective on the Taguchi Methods," Quality Progress, June, pp. 44–52.

Gupta, Y. and Chow, W. S. (1985), "Twenty-Five Years of Life Cycle Costing - Theory and Applications: A Survey," *International Journal of Quality & Reliability Management*, Vol. 2, No. 3, pp. 51–76.

Kackar, Raghu. (1985), "Off-Line Quality Control, Parameter Design, and the Taguchi Method," *Journal of Quality Technology*, Vol. 17, No. 4, pp. 176–188.

Kohoutek, H. J. (1996), "Economics of Reliability," in *Handbook of Reliability Engineering and Management*, Second Edition, McGraw-Hill, NY.

Logothetis, N. and Salmon, J. P. (1988), "Tolerance Design and Analysis of Audio-Circuits," *Taguchi Methods: Proceedings of the 1988 European Conference*, Elsevier Applied Science, London, England, pp. 161–175.

Meisl, C. J. (1990), "Parametric Cost Analysis in the TQM Environment," *Paper presented at the 12th Annual Conference of International Society of Parametric Analysts*, San Diego, CA.

Phadke, S. M. (1989), *Quality Engineering Using Robust Design*, Prentice Hall, Englewood Cliffs, NJ.

Seldon, M. R. (1979), *Life Cycle Costing: A Better Method of Government Procurement*, Westview Press, Boulder, CO.

Stoecker, W. F. (1989), *Design of Thermal Systems*, Third Edition, McGraw Hill, New York, NY, pp. 149–152.

Sullivan, L. P. (1987), "The Power of Taguchi Methods," Quality Progress, June, pp. 76–79.

Taguchi, G. (1986), *Introduction to Quality Engineering*, Asian Productivity Organization, Distributed by American Supplier Institute Inc., Dearborn, MI.

Taguchi, G., Elsayed, E. and Hsiang, T. (1989), *Quality Engineering in Production Systems*, McGraw Hill, New York, NY.

Taguchi, G. and Konishi, S. (1987), *Orthogonal Arrays and Linear Graphs*, American Supplier Institute Inc., Dearborn, MI.

Wang, J. X. (1991), "Fault Tree Diagnosis Based on Shannon Entropy," *Reliability Engineering and System Safety*, Vol. 34, pp. 143–167.

Wang, J. X. (1996), "Complexity as a Measure of the Difficulty of System Diagnosis," *International Journal of General Systems*, Vol. 24, No. 3, pp. 257–269.

Wang, J. X. (2002), *What Every Engineer Should Know about Decision Making under Uncertainty*, CRC Press, Boca Raton, FL.

Wang, J. X. (2008), *What Every Engineer Should Know about Business Communication*, CRC Press, Boca Raton, FL.

Wang, J. X. (2010), *Lean Manufacturing Business Bottom-Line Based*, CRC Press, Boca Raton, FL.

Wang, J. X. (2015), *Cellular Manufacturing Mitigating Risk and Uncertainty*, CRC Press, Boca Raton, FL.

Wang, J. X. (2017), *Industrial Design Engineering: Inventive Problem Solving*, CRC Press, Boca Raton, FL.

Wang, J. X. (2019), "Complexity as a Measure of the Difficulty of System Diagnosis in Next Generation Aircraft Health Monitoring System," SAE Technical Paper 2019-01-1357, doi:10.4271/2019-01-1357.

Wang, J. X. (2019), "A Dynamic Fault Tree Approach for Time-Dependent Logical Modelling of Autonomous Flight Systems," SAE Technical Paper 2019-01-1358, doi:10.4271/2019-01-1358.

Wille, R. (1990), "Landing Gear Weight Optimization Using Taguchi Analysis," *Paper presented at the 49th Annual International Conference of Society of Allied Weight Engineers Inc.*, Chandler, AR.

Wolfe, M. G., Oliver, M. B. and McClain, C. J. (1990), "System Modeling: Requirement for a New Generation of Cost Modeling," *Paper presented and distributed at the AIAA Symposium, Space Systems Cost Estimation*, Paper no: IAA-CESO-31 (90), San Diego, CA.

8 Schedule Risk
Identifying and Controlling Critical Paths

8.1 SCHEDULE: DELIVER ENGINEERING PRODUCTS ON TIME

The Great Wall in China, a huge engineering project 2,000 years ago, was developed for defense. The schedule of its design and construction was driven by the timing of the offense from another side of the wall. Engineers tried every means to integrate local defense systems into a successful military construction before another army launched their offense.

"An essential part of a plan is implementation". A good project plan that is never implemented is no better than no plan; a mediocre project plan, well-implemented, may have more impact than a great plan poorly implemented. The engineers of the Great Wall understood the risk of not meeting schedule and developed a design that could be implemented within the project schedule.

Managing schedule risk means arranging the project tasks – and their resources (human and other) – in a sequence that facilitates their completion. Often, this is done by setting "milestones" to gauge one's progress. Sometimes, it is important not only to accomplish the project, but to do so with the minimum resources possible. Managing schedule risk usually requires breaking the project into component tasks, figuring out the order in which the tasks must be completed and assigning the necessary resources to each task.

Managing schedule risk is an important part of project management. The simplest project management tool was developed by Henry Gantt and bears his name. Gantt Charts plot the steps necessary to complete a project against a timeline. Each task or activity is plotted by a bar or line which begins at a definite time and ends at a defined time. The activity bars are arranged in ascending or descending sequence of time (i.e., in the order in which the tasks begin or the order in which they end). The Gantt Chart provides milestones, or markers, for assuring that a program is on track.

While Gantt Charts have the advantage of simplicity, and do perform the work of displaying the tasks and some of the order in which they are to be performed, they can hide as much as they display. The linkages between the various tasks are not made explicit nor are the relationship between resources and the time to complete a task. While Gantt Charts can tell a manager whether a program is on track, they often cannot provide information about how to get it back on track.

Project Evaluation and Review Technique (PERT) was first used in the 1950s on projects like the construction of the Polaris submarine – a project which required coordinating the activities of 250 contractors and 9,000 subcontracts. Like Gantt Charts, PERT requires task definition and duration; in addition, PERT requires

DOI: 10.1201/9781003371014-8

specification of the relationship between tasks and the resources required to complete each task in a given period of time. Once the network of activities is determined, the Critical Path Method (CPM) is used to reallocate resources to find the shortest time in which the project can be completed (and what resources would be needed to do that).

Take a simple example: suppose you plan to build an addition on your house. First, you have to decide what you are going to do. For this example, you might have to build a foundation, frame the addition, do the electrical work, sheet-rock and trim, roofing, landscaping, exterior painting, and interior painting/finishing. Then you'd have to decide in what order things have to be done. For example, you can't frame the addition until the foundation is done, and you can't put on the roof until the framing is done. However, once the framing is done, the electrician can come in (but the sheet-rock can't go up until after that, and the interior painting is last) and the exterior painting and the landscaping can be done at the same time. Then you'd have to estimate how long it will take to do each task. The foundation could be dug, poured, and set in, say, two weeks – or one week if you really pushed, three weeks if you get bad weather. Then you could add up the time it takes to do the various tasks to estimate how long it should take to complete the project. You could even shorten the time it took to complete the project if you sped up tasks that were holding back other tasks. This is, in effect, a PERT. In a "picture", one might draw something like Figure 8.1.

Now, suppose you want to get this addition built as quickly as possible. You could always "crash" the project, using CPM. The idea is to use estimates of time and cost to find the most *efficient* use of resources to get the job done. This means getting it done in the least possible time, neither wasting nor sparing costs. To do this, you might develop a table of time and cost information for each task. What is the "normal" (expected) time the task should take, and what is the least ("crashed") time in which it could be done? And how much does it cost in each case for each task? Using this data, you could develop a "cost per week" for accelerating (the

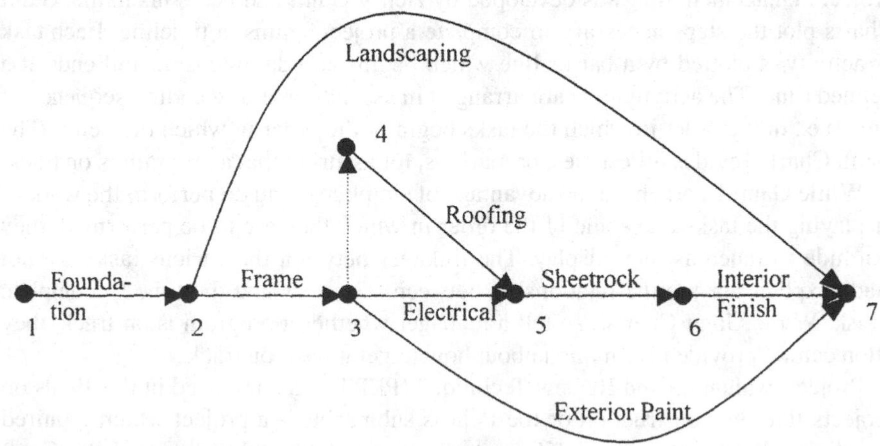

FIGURE 8.1 PERT Chart for house addition.

accelerated differential cost divided by the time saved). Using your PERT Chart, you could accelerate the cheapest activity path that is on the longest path for getting the job done (this is called the "critical path"). You could repeat this process, one path at a time, until all the paths have been optimally minimized. Then you should ease up on all the "non critical paths", just to the point that "all paths are critical" (i.e., so everything comes together at the same time – why spend more for shortening a task when it will not affect the project time?).

In the example above, you might develop a table that looks something like this:

Path	Name	Time Normal	Accelerated	Cost Normal	Accelerated
1-2	Foundation	2	1	$3,000	$4,000
2-3	Framing	4	1	$4,000	$8,000
3-4	Dummy				
3-5	Electrical	0.4	0.2	$1,000	$1,500
4-5	Roofing	1	0.4	$1,000	$1,750
5-6	Sheet-rock	1	0.4	$1,000	$1,500
6-7	Interior Finish	2	1	$1,500	$3,000
2-7	Landscaping	0.4	0.2	$500	$650
3-7	Exterior Paint	0.8	0.4	$600	$1,000

Using this information, you would expect to complete the addition in eight weeks at a cost of $12,600. If you pushed the project to the limit, you could get it done in a little over 3.8 weeks for $8,800 additional.

A simple project like this can be done fairly easily with paper and pencil. However, today's engineers would use a computer program such as Microsoft Project for Windows if they want to build another Great Wall.

8.2 CRITICAL PATH: DRIVER OF SCHEDULE RISK

Tools for analyzing schedule risk make a few assumptions, and use only a few technical terms. PERT and CPM assume that all activities have distinct beginning and ending points. In the earlier example of constructing an addition to a house, the landscaping activity might not need to be a distinct activity – as time is available, workers could push the fill back around the foundation, police the site, and grade the soil. But PERT need not take those random opportunities into account.

PERT and CPM also assume that the estimates of the time needed can be well-defined mathematically. It takes a fair bit of experience to accurately gauge the time a task will take. Barring such experience, time estimates will be highly uncertain and the quantitative measures of "normal" time and the probability estimates of completion time will be highly unreliable.

Third, resources must be able to be shifted from one activity to another. In the housing addition example, a carpenter's slack time cannot be shifted to an electrician's job or vice versa (the Trades and the Building Inspector would not approve). However, the *dollars* spent for hiring labor *can* be shifted from carpenters to electricians fairly easily. It depends on whether you are scheduling your work crews or hiring workers whether or not (in this example) you can shift resources.

Fourth, quantitative analysis in PERT and CPM assume that cost is a direct function of time and that those costs are evenly spread over the time (i.e., there is no "lumpiness" of activities). For example, the cost of hanging sheet-rock is mostly in the early stages of screwing the sheet-rock to the studs and the original taping. The subsequent sanding and additional layers of mud take a relatively short time, but require at least a day for drying between each application of mud. There is no cost to this drying time, but it cannot be shortened. In cases like these, fractional increases or decreases in time may not make sense.

Finally, PERT and CPM assume that the time value of money is not an issue. In other words, there is no need to consider discounting of future costs (or benefits), nor are there indirect monetary benefits to time (other than those incorporated in the costs to "crash" the time). For most projects, these assumptions are acceptable. For really large-scale projects (like the Polaris submarine project or even the construction of, say, a civic center), the model can be adjusted to include discounting or (for example) the time value of money as reflected in the construction-mortgage rate.

There are also a few technical terms used in PERT and CPM:

- **Critical Path:** the "longest" path (in terms of time) to the completion of a project. The critical path is the work-path which, if shortened, would shorten the time it takes to complete the project. Activities off the critical path would not affect completion time even if they were done more quickly.
- **Slack Time:** the margin between when a task is expected to be completed and the time when it must be completed if the project is to be done on time. Slack time is the difference between the expected time for arriving at the end of a task and the latest allowable time for finishing it.
- **Crashing:** shifting resources to reduce slack time so the critical path is as short as possible.
- **Dummy Variable:** each activity must be uniquely defined by its beginning and ending points. When two activities begin and end at the same time, a dummy variable (an activity which begins and ends at the same time) is inserted into the model to distinguish the two activities. In the housing-addition example, roofing and electrical work were going on at the same time, so a dummy variable was inserted to distinguish the two activities. While there is no intuitive need for dummy variables, they are needed to provide a completely specified, mathematical model of the PERT network.

The diagram in Figure 8.2 illustrates the kinds of situations we can represent in network diagrams.

In the last situation shown above, we have a dotted arc. This dotted line represents a *dummy activity.* Dummy activities often have a zero completion time and are used to represent precedence relationships that cannot be easily (if at all) represented using the actual activities involved in the project.

For example for the house addition project shown in Figure 8.1, the activity "Frame" must be finished before the activity "Roofing" can start; the easiest way

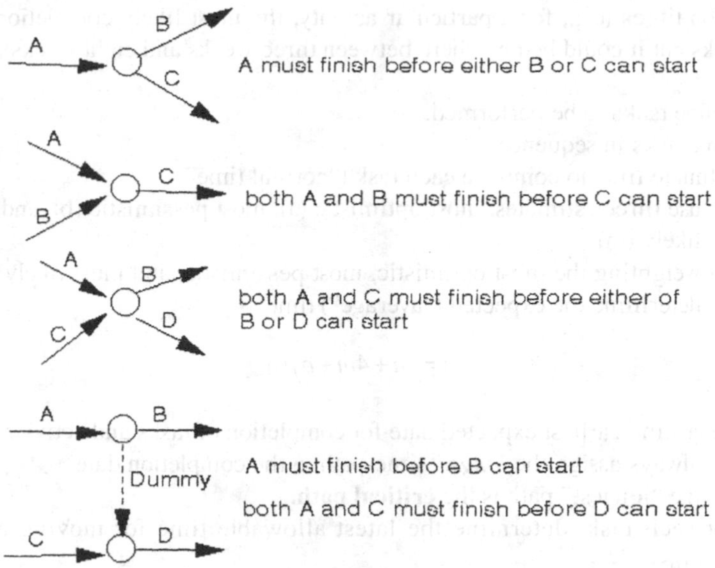
A
B
C
A must finish before either B or C can start
A
B
C
both A and B must finish before C can start
A
B
C
D
both A and C must finish before either of
B or D can start
A
B
Dummy
C
D
A must finish before B can start
both A and C must finish before D can start

FIGURE 8.2 Representation of activity precedence.

to represent this on the network diagram is by having a dummy activity, with a zero completion time, directed from node 3 (the end of activity "Frame") to node 4 (the start of activity "Roofing").

Dummy activities can have a non-zero completion time, e.g., if (taking the example mentioned in the previous paragraph) there must be a two-week delay between finishing activity "Frame" and starting activity "Roofing".

Often in drawing large networks, we find that the easiest way to represent some precedence relationships is by dummy activities. Once having drawn the network, it is a relatively easy matter to analyze it to find the critical path.

8.3 FIND AND ANALYZE CRITICAL PATH

Calculations for PERT and CPM are not particularly difficult – they require no more than simple arithmetic. They *can* get lengthy and sometimes convoluted, especially since CPM requires frequent recalculation. So, while it can be done with pencil and paper, frequently it is done with the aid of computers – simple spreadsheets, or even dedicated application programs like Microsoft Project.

When calculating PERT and CPM without a dedicated application program, it is best to approach it in a stepwise process (Krueckeberg & Silver, 1974).

8.3.1 PERT Analysis

PERT was developed by the U.S. Navy for the planning and control of the Polaris missile program and the emphasis was on completing the program in the shortest possible time. In addition, PERT had the ability to cope with uncertain activity completion times (e.g., for a particular activity, the most likely completion time is four weeks but it could be anywhere between three weeks and eight weeks).

1. Define tasks to be performed.
2. Link tasks in sequence.
3. Estimate time to complete each task ("normal time")
   - use three estimates: most optimistic (a), most pessimistic (b), and most likely (m)
   - weighting the most optimistic, most pessimistic, and most likely time, determine the expected ("average") time:

$$t = (a + 4m + b)/6$$

4. Determine earliest expected date for completion of tasks and activity
   - always assign the *longest* time-path to the completion date
   - the "longest" path is the **critical path.**
5. For each task, determine the latest allowable time for moving to the next task
   - the difference between latest time and expected time is the **slack time.**
6. Determine the standard deviation of meeting the expected time
   - use the time range (*b*–*a*) to estimate standard deviation of the time for each activity (i.e., estimate of average deviation from expected time):

$$sd = (b - a)/6$$

   - add the estimated deviations along the critical path to determine the spread of completing the project within a specified time:

$$\text{Standard Deviation} = \text{SqRt}\left(\text{Sum}\left[sd{**}2\right]\right)$$

   - read this formula as "Standard deviation equals the square root of the sum of the squared standard deviations for each activity on the critical path".
7. In PERT, shift the allocation of resources from slack activities to activities on the critical path, and revise time estimates and probability estimates. Usually, you would not settle just for shifting based on time saving, but would move at this point to CPM and consider time *and* money in determining the optimal balance.

A typical PERT table might have the following structure:

Activity	Beginning	Ending	A	M	B	Expected	*Sd*
Foundation	1	2	1	2	3	2	.33
Frame	2	3	1	4	6	4	.83

8.3.2 CPM Analysis

CPM was developed by du Pont and the emphasis was on the trade-off between the cost of the project and its overall completion time (e.g., for certain activities, it may be possible to decrease their completion times by spending more money – how does this affect the overall completion time of the project?).

1. Develop time and cost data ("normal" and "crashed") for all tasks.
2. Develop cost-per-week for crashing (difference in cost divided by time saved).
3. Develop project network (PERT).
4. Accelerate the activity *on the critical path* with the lowest cost-for-accelerating.
5. Recalculate the project network (the critical path might have been changed!).
6. Repeat steps 4 and 5 until all the paths have been minimized.
7. Ease up on all non-critical paths, just to the point that all paths are critical.

A typical CPM table might have the following structure:

Activity	Begin	End	Time-Crash	Time-Normal	Cost-Crash	Cost-Normal	Time Saved	Cost Increase	Cost/Week
Foundation	1	2	1	2	4,000	3,000	1	1,000	1,000
Frame	2	3	1	4	8,000	4,000	3	4,000	1,333

In addition to tabular data, both CPM and PERT will generally include a graphic presentation of the network of activities, usually with the length of each activity (in time) indicated and the critical path marked distinctively.

Interpreting PERT and CPM is fairly straightforward. They do not tell you what to do, but they provide the consequences in terms of time and resources (usually money) of alternative choices you might make, *given the assumptions you built into the analysis.*

It is generally a good idea to review the following five assumptions whenever you interpret the results of a program management analysis, particularly CPM:

1. All tasks have distinct beginning and ending points.
2. Time estimates are quantitatively well-defined.
3. Resources can be shifted from one task to another.
4. Cost of each activity is evenly spread over time.
5. No discounting or indirect costs.

The probability of meeting the expected time, which is part of PERT, also requires some explanation. By adding the standard deviation of the time estimate for each activity into a pooled estimate, you create the opportunity to estimate the probability of completing the project within a certain time. This is a crude estimate (*not a guarantee!*). If the pooled deviations are treated as confidence limits, then 95% of the time the project can be expected to be completed somewhere between ±2 times the standard deviation of the expected time, assuming the distribution of completion times is normally distributed. In other words, if the PERT analysis arrives at an expected completion time of ten weeks and the standard deviation of that estimate (i.e., the pooled standard deviations of the various tasks) is 1.0, then you can expect that 95% of the time (or, you can predict with 95% confidence), the project will be completed in 8–12 weeks. If you are more of a risk-taker, the confidence interval for ±1 standard deviation is about 68%.

8.4 SCHEDULE RISK FOR A SINGLE DOMINANT CRITICAL PATH

Schedule risk assessment is the general name given to certain specific techniques which can be used for the assessment, management, and control of projects. Typically, all engineering projects can be broken down into:

- separate activities (tasks) – where each activity has an associated duration or completion time (i.e., the time from the start of the activity to its finish);
- precedence relationships – which govern the order in which we may perform the activities, e.g., in a project concerned with building a house the activity "erect all four walls" must be finished before the activity "put roof on" can start.

The problem is to bring all these activities together in a coherent fashion to complete the project.

Activity networks, as described above, are often used in the planning and scheduling of projects, including construction, and research and development projects. In the case of activity networks, the branches between node points represent specific activities. Clearly, there may be certain activities that must be completed before subsequent ones can be started; hence, in an activity network, certain procedure logic must be observed.

As shown in Section 8.3, the CPM is often used to plan and schedule construction projects. This is based on the assumption that the duration of the individual activities in a network is deterministic (i.e., no uncertainty); thus, the completion time of the project is determined by the "critical path" which is the activity path with the longest duration.

However, as there are invariably uncertainties in the estimation of the duration of the individual activities, the project completion time may be evaluated only with an associated uncertainty. In this extension to the basic PERT and CPM, for each activity, not a single completion time but three times are required:

- optimistic time (t_1) – the completion time if all goes well;
- most likely (modal) time (t_2) – the completion time we would expect under normal conditions;
- pessimistic time (t_3) – the completion time if things go badly.

This use of three time estimates in the PERT technique was previously discussed (see Section 8.3).

These three times are combined into a single figure, the expected activity completion time, given by $(t_1 + 4t_2 + t_3)/6$. This figure is used as the activity completion time when carrying out the calculations presented before to find the project completion time and the critical activities.

Note that this weighting of the optimistic, most likely, and pessimistic times of 1/6:4/6:1/6 is essentially fixed and cannot be altered (as the underlying theory depends on these weights).

In addition, through the use of statistical probability distributions, we can get an idea of how this project completion time might vary (remember we no longer know the individual activity completion times with certainty).

Essentially, we can find answers to questions such as, what is the probability that:

- the project will take longer than…?
- the project will be finished by…?
- a particular activity will take longer than…?
- a particular activity will be finished by…?

We will illustrate schedule risk assessment with reference to the following example: suppose that we are going to carry out a redesign of a product and its associated packaging. We intend to test market this redesigned product and then revise it in the light of the engineering test results, finally presenting the results to the management of the company.

The key question is: *How long will it take to complete this project?*

After much *thought*, we have identified the following list of separate activities together with their associated completion times (see Table 8.1). Aside from this list of

TABLE 8.1
Activities for Product Redesign and Packaging

Activity Number (i)	Activity Name	Optimistic Time (TO_i)	Most likely Time (T_i)	Pessimistic Time (TP_i)	Standard Deviation $\sigma_i = (TP_1 - TO_i)/6$
1	Redesign product	3	6	12	1.5
2	Redesign packaging	1	2	4	0.5
3	Order and receive components for re-designed product	1	3	6	0.75
4	Order and receive material for redes-igned packaging	1	2	4	0.5
5	Assemble products	2	4	8	1
6	Makeup packaging	0.5	1	2	0.25
7	Package redesigned product	0.5	1	2	0.25
8	Test redesigned product	3	6	12	1.5
9	Revise redesigned product	1.5	3	6	0.75
10	Revise redesigned packaging	0.5	1	2	0.25
11	Present results to the management	0.5	1	2	0.25

activities, we must also prepare a list of precedence relationships indicating activities which, because of the logic of the situation, must be finished before other activities can start (e.g., in the above list, activity number 1 must be finished before activity number 3 can start).

It is important to note that, for clarity, we try to keep this list to a minimum by specifying only *immediate* relationships, that is relationships involving activities that "occur near to each other in time".

For example, it is plain that activity 1 must be finished before activity 9 can start but these two activities can hardly be said to have an immediate relationship (since many other activities after activity 1 need to be finished before we can start activity 9).

Activities 8 and 9 would be examples of activities that have an immediate relationship (activity 8 must be finished before activity 9 can start). Note here that specifying non-immediate relationships merely complicates the calculations that need to be done – it does not affect the final result. Note too that, in the real world, the consequences of leaving out precedence relationships are much more serious than the consequences of including unnecessary (non-immediate) relationships.

Aided with this list of the activities presented in a logical/chronological order, the following is a list of immediate precedence relationships (see Table 8.2).

For each activity, in turn, the key to constructing this table is to ask the question: "*What activities must be finished before this activity can start?*" Note here that:

- Activities 1 and 2 do not appear in the right-hand column of the above table. This is because there are no activities which must finish before they can start (i.e., both activities 1 and 2 can start immediately).
- Two activities (5 and 6) must be finished before activity 7 can start.
- It is plain from this table that non-immediate precedence relationships (e.g., "activity 1 must be finished before activity 9 can start") need not be included in the list since they can be deduced from the relationships already in the list.

TABLE 8.2
Precedence Relationship for Product Redesign and Packaging

Preceding Activity Number		Following Activity Number	
1	Must be finished before	3	Can start
2		4	
3		5	
4		6	
5, 6		7	
7		8	
8		9	
8		10	
9, 10		11	

Once we have completed our list of activities and our list of precedence relationships, we combine them into a diagram/picture (called a *network* – which is where the name network analysis comes from). We do this below.

Note first however that we asked the key question above: *How long will it take to complete this project?* (i.e., complete all the activities while respecting the precedence relationships). One answer could be if we first do activity 1, then activity 2, then activity 3,..., then activity 10, then activity 11. Such an arrangement would be possible here (check the precedence relationships above), and the project would then take the sum of the activity completion times, or 30 weeks.

However, could we do the project in less time? It is clear that logically we need to amend our key question to be: *What is the minimum possible time in which we can complete this project?*

We shall see below how the network analysis diagram/picture we construct helps us to answer this question.

In constructing the network as shown in Figure 8.3, we use the precedence relationships to construct it from left to right – adding activities (arcs) to the network as the precedence relationships indicate that activities can start. The key to constructing the network is the question: *"What activities can start now?"*

Initially, for example, both activities 1 and 2 can start. Once activity 1 has been finished activity 3 can start, once activity 2 has been finished activity 4 can start, etc. The nodes of the network are numbered 1,2,3,..., etc. Note:

- that all arcs have arrows attached to them (indicating the direction the project is flowing in);
- the way the relationship "activities 5 and 6 must be finished before activity 7 can start" is represented.

The PERT analysis for this example is shown in Table 8.3. Note here that one of the columns in that figure corresponds to the expected activity completion time $(t_1 + 4t_2 + t_3)/6$.

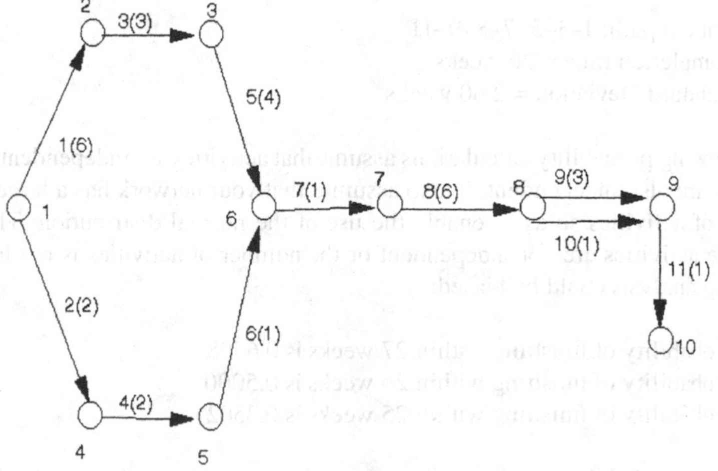

FIGURE 8.3 Network for product redesign and packaging.

TABLE 8.3
PERT Analysis for Network Problem

Activity Number	Exp. Time	Variance	Earliest Start	Latest Start	Earliest Finish	Latest Finish	Slack LS-ES
1	6.5	2.25	0	0	6.5	6.5	Critical
2	2.17	0.25	0	8.67	2.17	10.83	8.67
3	3.25	0.56	6.5	6.5	9.75	9.75	Critical
4	2.17	0.25	2.17	10.83	4.33	13	8.67
5	4.33	1.00	9.75	9.75	14.08	14.08	Critical
6	1.08	0.0625	4.33	13.0	5.41	14.08	8.67
7	1.08	0.0625	14.08	14.08	15.17	15.17	Critical
8	6.5	2.25	15.17	15.17	21.67	21.67	Critical
9	3.25	0.5625	21.67	21.67	24.92	24.92	Critical
10	1.08	0.0625	21.67	23.83	24.92	24.92	2.17
11	1.08	0.0625	24.92	24.92	26	26	Critical

The critical path is found to be 1–3–5–7–8–9–11. From Table 8.3, the expected estimates for project completion time can be calculated by

$$\text{Completion Time} = T_1 + T_3 + T_5 + T_7 + T_8 + T_9 + T_n$$
$$= 6.5 + 3.25 + 4.33 + 1.08 + 6.5 + 3.25 + 1.08$$
$$= 26(\text{weeks})$$

The standard deviation, σ, can be calculated by

$$\alpha = \sqrt{\alpha_1^2 + \alpha_3^2 + \alpha_5^2 + \alpha_7^2 + \alpha_8^2 + \alpha_9^2 + \alpha_{11}^2} = 2.60$$

- Critical path: 1–3–5–7–8–9–11
- Completion time = 26 weeks
- Standard Deviation = 2.60 weeks

The following probability calculations assume that activities are independent and that all paths are also independent. It also assumes that your network has a large enough number of activities so as to enable the use of the normal distribution. Therefore, when the activities are not independent or the number of activities is not large, the following analysis could be biased:

- Probability of finishing within 27 weeks is 0.6498
- Probability of finishing within 26 weeks is 0.5000
- Probability of finishing within 25 weeks is 0.3502

Using this probability analysis, it is possible to construct the graph shown in Figure 8.4, for a range of possible project completion times. We have plotted the

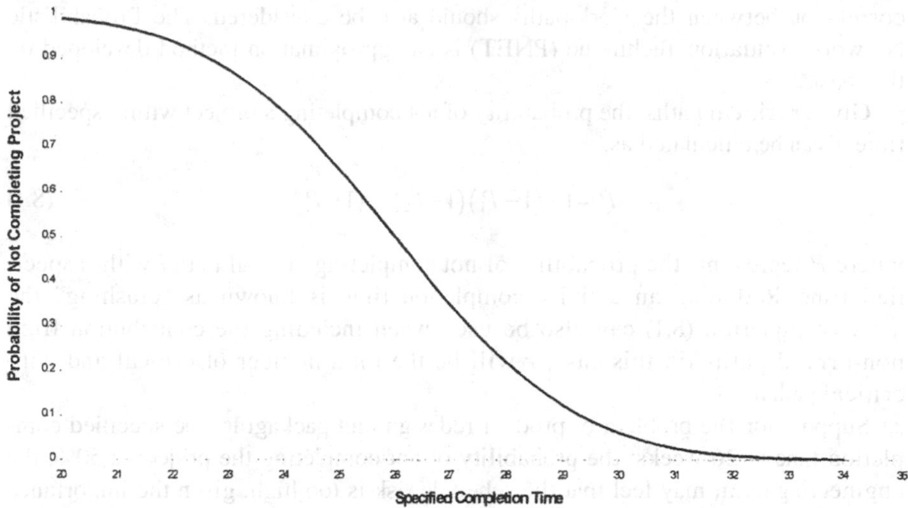

FIGURE 8.4 Probability of not completing project within specified project completion time.

probability that the project will have been completed by that time. This plot is based on the assumption that one critical path is dominant. The analysis for multiple critical paths will be presented in Section 8.5.

Note here that the curve shown below is not fixed – we may be able to change its shape by refining the time estimates (optimistic, most likely, pessimistic) we used which lead to the curve (i.e., to reduce the uncertainty associated with each activity, and hence reduce the uncertainty associated with the completion of the entire project).

Computer packages are widely available for schedule risk assessment (including packages on Personal Computers). Typically, such packages will:

- draw network diagrams;
- calculate critical activities, overall project completion time;
- cope with uncertain activity times;
- perform project time/project cost trade-off;
- deal with resource allocation/resource smoothing problems;
- deal with multiple projects;
- provide facilities for updating the network as the project progresses.

8.5 SCHEDULE RISK FOR MULTIPLE CRITICAL PATHS

Using CPM, the probability is determined solely on the basis of the critical path, which is also the basis of the method of PERT. This approximation would be reasonable when there is a dominant critical path; however, results obtained on this basis are invariably on the optimistic side, especially when multiple potential critical paths exist.

To obtain a better estimate of the schedule risk, the multiple work-paths and their impact on project schedule must obviously be included; nevertheless, the mutual correlation between the work-paths should also be considered. The Probabilistic Network Evaluation Technique (PNET) is an approximation method developed on this basis.

Given r critical paths, the probability of not completing a project within specified time T can be calculated as:

$$P = 1 - (1 - P_1)(1 - P_2)\ldots(1 - P_r) \tag{8.1}$$

where P_i represents the probability of not completing critical path i within specified time. Reducing an activity completion time is known as "crashing" the activity. Equation (8.1) can also be used when including the contribution from non-critical paths (in this case, r will be the total number of critical and non-critical paths).

Suppose for the problem of product redesign and packaging, the specified completion time is 26 weeks; the probability of not completing the project is 50%; the engineering team may feel that the schedule risk is too high, given the importance of the project. They shorten the duration of the "revising redesigned products" activity number 9 to one day by working three shifts on that day. This will result in the additional overtime cost of $500. The revised activity list and network are shown in Figure 8.5 and Table 8.4.

As shown in Table 8.5, there are two critical paths after incorporating the crash activity:

Critical Path 1: 1–3–5–7–8–9–11
Critical Path 2: 1–3–5–7–8–10–11

- Critical path: 1–3–5–7–8–9–11 and 1–3–5–7–8–10–11
- Completion time = 24 weeks
- Standard Deviation = 2.50

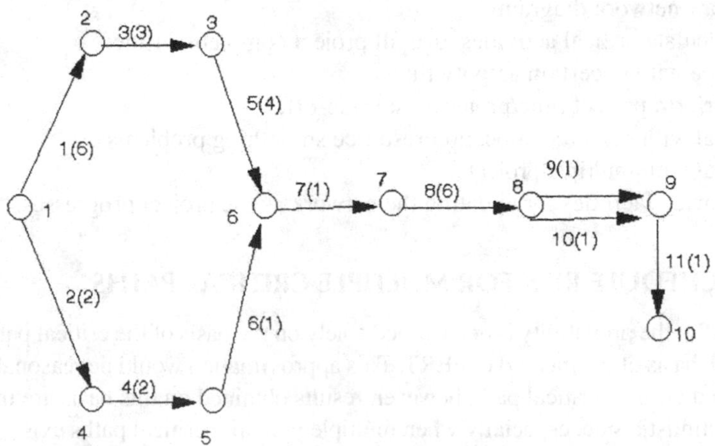

FIGURE 8.5 Network for product redesign and packaging after "crash".

TABLE 8.4
Revised Table of Activities for Product Redesign and Packaging

Activity Number (I)	Activity Name	Optimistic Time (TO_i)	Most Likely Time (T_i)	Pessimistic Time (TP_i)	Standard Deviation $\sigma_i = TP_i - TO_i)/6$
1	Redesign product	3	6	12	1.5
2	Redesign packaging	1	2	4	0.5
3	Order and receive components for redesigned product	1	3	6	0.75
4	Order and receive material for redesigned packaging	1	2	4	0.5
5	Assemble products	2	4	8	1
6	Make up packaging	0.5	1	2	0.25
7	Package redesigned product	0.5	1	2	0.25
8	Test redesigned Product	3	6	12	1.5
9	Revise redesigned Product	0.5	1	2	0.25
10	Revise redesigned packaging	0.5	1	2	0.25
11	Present results to the management	0.5	1	2	0.25

For critical path 1,

$$\begin{aligned}\text{Completion Time } T_{C1} &= T_1 + T_3 + T_5 + T_7 + T_8 + T_9 + T_{11} \\ &= 6.5 + 3.25 + 4.33 + 1.08 + 6.5 + 1.08 + 1.08 \\ &= 24(\text{weeks})\end{aligned}$$

TABLE 8.5
PERT Analysis for the Revised Network Problem

Activity Number	Exp. Time	Variance	Earliest Start	Latest Start	Earliest Finish	Latest Finish	Slack LS-ES
1	6.5	2.25	0	0	6.5	6.5	Critical
2	2.17	0.25	0	8.67	2.17	10.83	8.67
3	3.25	0.56	6.5	6.5	9.75	9.75	Critical
4	2.17	0.25	2.17	10.83	4.33	13	8.67
5	4.33	1.00	9.75	9.75	14.08	14.08	Critical
6	1.08	0.0625	4.33	13.0	5.41	14.08	8.67
7	1.08	0.0625	14.08	14.08	15.17	15.17	Critical
8	6.5	2.25	15.17	15.17	21.67	21.67	Critical
9	1.08	0.0625	21.67	21.67	22.75	22.75	Critical
10	1.08	0.0625	21.67	21.67	22.75	22.75	Critical
11	1.08	0.0625	22.75	22.75	23.83	23.83	Critical

The standard deviation can be calculated by

$$\alpha_{C1} = \sqrt{\alpha_1^2 + \alpha_3^2 + \alpha_5^2 + \alpha_7^2 + \alpha_8^2 + \alpha_9^2 + \alpha_{11}^2} = 2.50$$

For critical path 2,

$$\begin{aligned}\text{Completion Time } T_{C2} &= T_1 + T_3 + T_5 + T_7 + T_8 + T_{10} + T_{11} \\ &= 6.5 + 3.25 + 4.33 + 1.08 + 6.5 + 1.08 + 1.08 \\ &= 24\,(\text{weeks})\end{aligned}$$

The standard deviation can be calculated by:

$$\alpha_{C2} = \sqrt{\alpha_1^2 + \alpha_3^2 + \alpha_5^2 + \alpha_7^2 + \alpha_8^2 + \alpha_{10}^2 + \alpha_{11}^2} = 2.50$$

According to Equation (8.1), the probability of not completing project by time T can be calculated as follows:

$$\begin{aligned}P &= 1 - (1 - P_1)(1 - P_2) \\ &= 1 - \{\Phi[(T - 24)/2.5]\}^2\end{aligned}$$

As shown in Figure 8.6, the probability of not completing the project within 26 weeks has been reduced to 38% by the crash activity.

It can be profitable to capture the benefits that using schedule risk assessment can bring to a project.

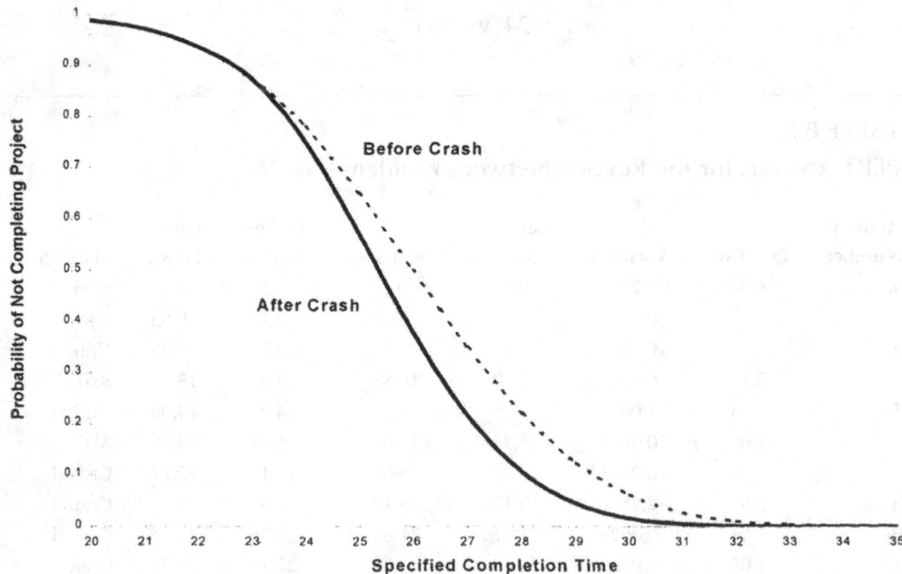

FIGURE 8.6 Probability of not completing project as reduced by crash activity.

8.5.1 Structure

Forming the list of activities, precedence relationships, and activity completion times causes a structuring of thought about the project and clearly indicates the separate activities that we are going to need to undertake, their relationship to one another, and how long each activity will take. Hence, network analysis is useful at the planning stage of the project.

8.5.2 Management

Once the project has started, the basic idea is that we focus management attention on the critical activities (because if these are delayed, the entire project is likely to be delayed).

In addition, for the non-critical activities, we have a natural ranking in terms of their slack (float) time. Plainly, activities with a smaller slack time rate more attention than those with a larger slack time.

It is relatively easy to update the network, at regular intervals, with details of any activities that have been finished, revised activity completion times, new activities added to the network, changes in precedence relationships, etc., and recalculate the overall project completion time. This gives us an important engineering tool for managing (controlling) the project.

Plainly, it is also possible to ask (and answer) "what if" questions relatively easily (e.g., what if a particular activity takes twice as long as expected – how will this affect the overall project completion time?).

The ability to tackle resource allocation and resource smoothing problems is also a great help in making efficient use of the resources available to the project manager.

It is also possible to identify activities that, at the start of the project, were non-critical but which, as the project progresses, approach the status of being critical. This enables the project manager to "head off" any crisis that might be caused by suddenly finding that a previously neglected activity has become critical.

8.6 PROBABILISTIC CRITICAL PATHS: QUANTIFYING COMPLEXITY OF PROJECT SCHEDULING

8.6.1 Probabilistic Critical Path

Using a three-point estimating method, where the three points can be acquired in different ways:

- Historical data
- Professional judgment
- Paired comparison selection

The 10%/90% points on a triangle distribution, which should accurately represent the task duration bounds, can be used as the starting point for a Monte Carlo simulation of the project network. As shown by Figure 8.7, a probabilistic critical path

POSSIBLE EFFECTS OF VARIANCE IN DETERMINING THE CRITICAL PATH

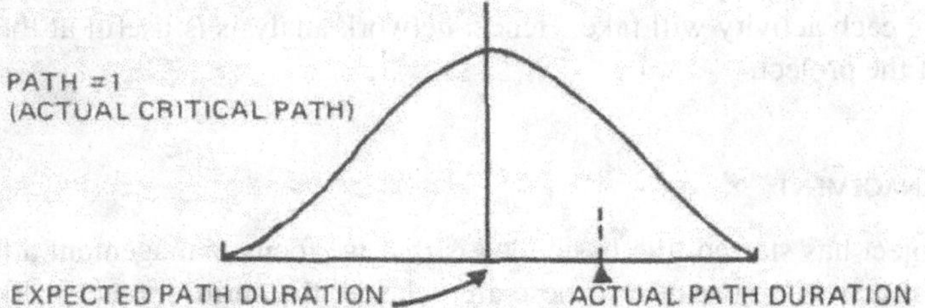

Probabilistic critical path: identify the driver for scheduling risk under uncertainty

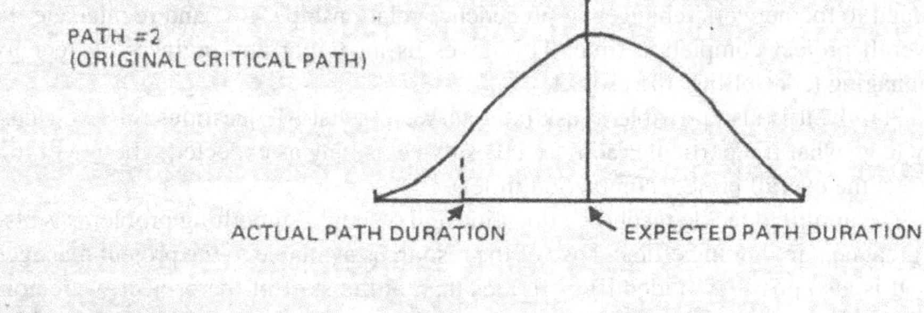

FIGURE 8.7 Probabilistic critical path: Identify the driver for scheduling risk under uncertainty.

can be found using these upper and lower bounds, a probability distribution, and the relevant data.

A list of tasks that are not on the critical path in the deterministic schedule is often provided by the tool used to analyze schedule risk. They show up on the critical path in the Monte Carlo simulation occasionally, typically 75% or more of the time. Following that, these tasks are gathered for examination using the PERT.

8.6.2 Identifying the Probabilistic Critical Path(s)

The schedule risk analysis (SRA) should be used to identify the sensitivity of the uncertainty and risks to the overall critical path calculations. The path with the highest cumulative impact from uncertainty and risks is likely to end up being the longest path when the effort is completed, according to an analysis of probabilistic critical paths. The SRA results will show the probability that tasks will develop into critical path activities. From this information, the probabilistic critical path can be better understood. Higher probabilities indicate a higher likelihood of a particular task becoming critical, or in other words, ending up on the critical path.

The likely critical path may, in many cases, change from the deterministic critical path to something else when uncertainty and risks are taken into account. It is important to compare the crucial activities list to both the deterministic critical path and the probabilistic critical path(s).

- These tasks might be reflected in the critical path report, presuming that the list of "critical activities" was taken into consideration as uncertainty distributions were developed.
- To fully understand why specific items on the project's list are not showing up as critical as a result of the SRA, any differences between the criticality path report and the project's "critical activities" list should be examined (for example, were the applied uncertainty bounds inadequate, are the items really not as "critical" to the schedule as initially thought, etc.).
- The network logic and durations for any unexpected activities that appear on the probabilistic critical path should be checked to make sure no mistakes were made.

In summary, making risk-informed decisions on the project's most critical tasks and critical path is enabled by performing SRA. It is strongly recommended to use a probabilistic schedule risk analysis (PSRA) as the basis for deciding where to place an adequate schedule margin. The PSRA results may be taken into account in addition to other variables, expert opinion, and general guidelines that frequently affect how much schedule margin should be applied to a timetable. Thorough examination of the deterministic and probabilistic critical paths should be used to decide where the appropriate margins should be placed. This more comprehensive method of comprehending how the project's probabilistic critical path develops as a result of taking into account task durations and sequencing, as well as schedule risks and uncertainties at any given time, provides a solid basis of estimate for not only establishing up and tracking margin, but also actively managing margin activities.

8.6.3 What Can We Learn from the Probabilistic Critical Path?

Where to prepare for schedule margin and risk reduction actions is the key piece of information. It gives insight into the possible risks associated with the plan as it is currently implemented. The plan is merely a means of looking into the future for a potential solution, one that is developing over time. It is a plan and assess, not a plan and forget. A complex program with significant levels of technical and programmatic uncertainty might benefit greatly from a probabilistic assessment of the plan, its cost, and its programmatic and technical risks.

BIBLIOGRAPHY

Ahuja, H. N., Dozzi, S. P. and Abourizk, S. M. (1994), *Project Management*, Second Edition, John Wiley & Sons, Inc., New York.

Ang, A. H-S. and Tang, W. H. (1984), *Probability Concepts in Engineering Planning and Design, Volume II - Decision, Risk, and Reliability*, John Wiley & Sons, New York.

Johnson, R. A. (1994), *Miller & Freund's Probability & Statistics for Engineers*, Fifth Edition, Prentice Hall, NJ.

Krueckeberg, D. A. and Silver, D. A. (1974), "Program Scheduling," *Urban Planning Analysis: Methods and Models*, John Wiley & Sons, Inc., NY, pp. 231–255.

Levin, R.I. and Kirkpatrick, C. A. (1966), *Planning and Control with PERT/CPM*, McGraw-Hill, NY.

Wang, J. X. (1991), "Fault Tree Diagnosis Based on Shannon Entropy," *Reliability Engineering and System Safety*, Vol. 34, pp. 143–167.

Wang, J. X. (1996), "Complexity as a Measure of the Difficulty of System Diagnosis," *International Journal of General Systems*, Vol. 24, No. 3, pp. 257–269.

Wang, J. X. (2002), *What Every Engineer Should Know about Decision Making under Uncertainty*, CRC Press, Boca Raton, FL.

Wang, J. X. (2008), *What Every Engineer Should Know about Business Communication*, CRC Press, Boca Raton, FL.

Wang, J. X. (2010), *Lean Manufacturing Business Bottom-Line Based*, CRC Press, Boca Raton, FL.

Wang, J. X. (2015), *Cellular Manufacturing Mitigating Risk and Uncertainty*, CRC Press, Boca Raton, FL.

Wang, J. X. (2017), *Industrial Design Engineering: Inventive Problem Solving*, CRC Press, Boca Raton, FL.

Wang, J. X. (2019), "Complexity as a Measure of the Difficulty of System Diagnosis in Next Generation Aircraft Health Monitoring System," SAE Technical Paper 2019-01-1357, doi:10.4271/2019-01-1357.

Wang, J. X. (2019), "A Dynamic Fault Tree Approach for Time-Dependent Logical Modeling of Autonomous Flight Systems," SAE Technical Paper 2019-01-1358, doi:10.4271/2019-01-1358.

9 Integrated Risk Management and Computer Simulation

9.1 AN INTEGRATED VIEW OF RISK

Risk means many things to many people. The dictionary defines risk as "exposure to the chance of injury or loss". In terms of insurance, it defines risk as "the hazard or chance of loss". Rowe (1994) defines risk as the downside of a gamble, which is described in terms of probability. Although there have been debates about the proper definition for risk, there is a general agreement that risk always involves two characteristics:

- Uncertainty – The event that characterizes the risk may or may not happen; i.e., there is no 100% probable risk;
- Loss – If the risk becomes a reality, unwanted consequences or losses will occur.

Kaplan and Garrick (1981) provide an excellent introduction to risk, both as a personal thing and as a mathematical entity. They note that risk is "relative to the observer" and has to do with both uncertainty and damage. They provide a first-level mathematical definition of risk as a "the set of triplets: $R = \{(S_i, P_i, X_i)\}$… where Si is a scenario identification or description, Pi is the probability of that scenario, and Xi is the consequence or evaluation measure of that scenario, i.e., the measure of damage". They further note that risk is multidimensional, i.e., there are many possible damage measures. Finally, they extend the definition of risk to a probabilistic family of multidimensional curves, which include the first-level definition of risk. For the mathematically inclined, very interesting concepts are to be found here.

Risk is the perceived extent of possible loss. Risk is individual to a person or organization because what is perceived by one as a major risk may be perceived by another as a minor risk. Perception is very much a factor. To some, risk is defined as the chance of loss. But in reality, it is more. Risk also concerns how much could be lost. Probabilistic risk analysis can define risk as the expected loss, or even as a family of probability curves. This has to do not just with the chance of loss, but more specifically with the extent of loss.

Risk concerns future happenings and opportunities. Today and yesterday are beyond the active concerns of risk engineering and management, as we are already reaping what was previously sowed by past actions. The question is, can we, by changing our actions today, create an opportunity for a different and hopefully better situation for tomorrow. Risk involves change, such as changes of design, manufacturing

DOI: 10.1201/9781003371014-9

process, engineering team, and management structure. Risk involves choice and the uncertainty that choice itself entails.

But what does risk have to do with competitiveness of engineering products? Everything! Each engineering decision has the possibility of resulting in loss. Each decision to introduce a new engineering product into the marketplace can result in varying degrees of loss or gain. To be an engineer is to accept risk, that is, the possibility of loss. A good engineering forte, however, is to make decisions that maximize possible gain, and hence minimize possible loss. This has to do with risk engineering and management. Risk engineering has to do with the engineering process used to estimate the extent of possible loss and gain of each engineering alternative. So does risk management which provides rationale for the decision-maker.

The risk associated with designing for competitive advantage can be decomposed into its primary components. This risk can be estimated in terms of the extent by which final cost is expected to exceed planned cost, the extent by which expected performance will fall short of technical specification, and the extent by which expected market introduction will fall short of meeting the planned introduction date. Note that risk, as estimated here, can be both positive and negative, is a dynamic measure over the life of a project, and is multidimensional. It should be eventually measured by integrating the cost risk, the quality risk, and the timeliness risk. Risk is the integrated effect of those things which went better than planned and those things which went worse than planned. The {expected value – planned value} is a projection of risk onto the basic dimensions of competitive advantage.

In this chapter, selected statistical and probability concepts are used to analyze the economic consequences of some decision situations involving risk and uncertainty and requiring engineering knowledge and input.

9.2 INTEGRATED RISK MANAGEMENT

Reactive risk strategies have been laughingly called the "Indiana Jones school of risk management" (Thomsett, 1992). In the movies that carried his name, Indiana Jones, when faced with overwhelming difficulty, would invariably say, "Don't worry, I'll think of something!" Never worrying about problems until they happened, Indy would react in some heroic way (see Figure 9.1).

Prior to the 1980s, government-funded projects followed a more-or-less regimented lifecycle. First, there was Program Initiation, followed by Concept Development, then Proof-of-Principle, then Full Scale Development, then Production/Acquisition, and finally Fielding/Deployment. Under this system, projects were allowed to fail. A program that failed to achieve substantial progress during, for example, Proof-of-Principle, was ended. The majority of engineering teams relied solely on reactive risk strategies.

At best, a reactive strategy monitors the project for likely risks. Resources are set aside to deal with them, should they become actual problems. More commonly, the engineering team does nothing about risks until something goes wrong. Then, the team flies into action in an attempt to correct the problem. This is often called the "fire-fighting mode". When this fails, "crisis management" takes over and the project is in real jeopardy.

FIGURE 9.1 Reactive risk strategy: "Don't worry. I'll think of something!".

During the 1980s, risk became so important that the previous lifecycle began to be ignored. Instead of ending an unsuccessful effort, the effort's sponsors and proponents began to retroactively define failures as "risk reduction efforts", based on the assumption that lessons learned would insure that a subsequent effort would be successful. This led some projects to actually have a Proof-of-Principle phase followed by a "Proof-of-Principle II" and even a "Proof-of-Principle III" phase.

A considerably more intelligent strategy for risk management is to be proactive. A proactive strategy begins long before technical work is initiated. Potential risks are identified, their probability and impact are assessed, and they are prioritized by importance. Then, the engineering team establishes a plan for managing risk. The primary objective is to avoid risk, but because not all risks can be avoided, the engineering team works to develop a contingency plan that will enable it to respond in a controlled and effective manner.

Attention to risk has affected the way governmental projects are structured and technologies are developed. Instead of initiating a project to develop a technology, which would involve assessing and acknowledging risk and suffering the various slings and arrows of the acquisition lifecycle, it is now common to use a "technology directorate" group to develop a technology, essentially replacing the old Concept Development phase with a "Technology Development Effort", managed with little public scrutiny and with fewer defined goals and restrictions, and no chance for "failure". Today, by the time a technology reaches the stage of becoming a formal R&D or acquisition program, there is little technical risk of failure. But there are still cost and schedule risks, and since money is crucial to everything, these are the risks that really matter.

Simply put, integrated risk management will focus management attention (and thus, resources) on the elements of a project that need such attention in order to permit the project to achieve its technical, schedule, and cost goals proactively. A secondary need for a risk analysis is to "prove" to one's superiors that the project is worthy of scarce funding because all risks have been taken into account and mitigated or minimized to some degree. Although secondary in importance, this last reason is the real reason that most risk analyses are performed. And this has had far-reaching impacts, changing the very way that projects are defined and executed.

The present worth (*PW*) method is based on the concept of equivalent worth of all cash flows relative to some base or beginning point in time called present. That is, all cash inflows and outflows are discounted to the present time at a minimum attractive rate of return (MARR), which is determined by the investment alternatives. For example, through leasing a computer, a computer service company gets an annual income of $650 for the next four years. Comparing with other potential usage of the computer system, the MARR is 11%. The *PW* of this leasing contract is

$$PW(11\%) = 650(1+11\%) + 650(1+11\%)^2 + 650(1+11\%)^3 + 650(1+11\%)^4$$
$$= \$2{,}016$$

Often engineers make decisions using economic principles when comparing different alternatives. These choices include the production method to use, the machinery to purchase, the number of people to hire, etc. As shown in Figure 9.2, a cash flow diagram provides an integrated view of technical, cost, and schedule risks over a product's useful life. The horizontal line is a time scale. Normally, years are given as

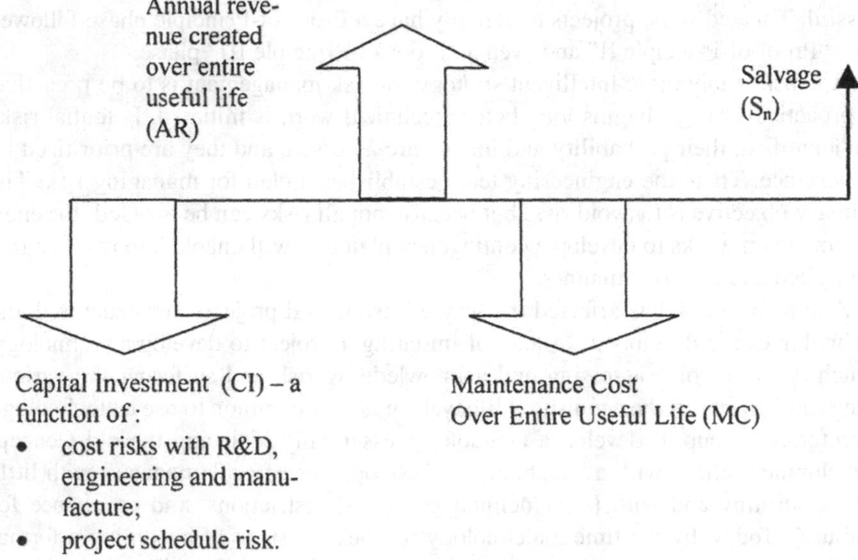

FIGURE 9.2 Cash flow diagram with an integrated view of cost, schedule, and technical risks.

the interval of time. The arrows signify cash flow. A downward arrow means money out and an upward arrow means money in.

Based on the integrated view of cost, schedule, and technical risks, the project *PW* can be calculated to evaluate engineering alternatives. As shown in Figure 9.2, three factors are used to compute these: capital investment, maintenance cost, and salvage.

Capital investment is influenced by project cost risk and schedule risk. Maintenance cost is the cost of annual repairs. Salvage is the net sum to be realized from the disposal of the system after service. For calculating the *PW*, all cash inflows and outflows are discounted to the present time at an interest rate that is generally the MARR.

$$PW = -CI + (AR - MC)\left[\frac{(1+i)^N - 1}{i(1+i)^N}\right] + \frac{S_N}{(1+i)^N} \tag{9.1}$$

where i = MARR and N is the product's useful life. The *PW* of an engineering alternative is a measure of how much money an individual or a firm could afford to pay for the project in excess of its cost. A positive *PW* for an engineering project is a dollar amount of profit over the minimum amount required by management. The probability of a negative *PW* denotes the integrated risk with cost, schedule, and technical issues. It is assumed that cash generated by the alternative is available for other uses that earn interest at a rate equal to the MARR.

The MARR is usually a policy issue resolved by the top management of an organization in view of numerous considerations, including:

- The amount of money available for investment, and the source and cost of these funds (i.e., equity funds or borrowed funds);
- The number of good projects available for investment and their purpose (i.e., whether they sustain present operations and are essential, or expand on present operations and are elective);
- The amount of perceived risk associated with investment opportunities available to the firm and the estimated cost of administering projects over short planning horizons vs. long planning horizons;
- The type of organization involved (i.e., government, public utility, or competitive industry).

A situation requiring decisions such as a design task, a new product, an improvement project, or a similar effort requiring engineering knowledge has two or more alternatives associated with it. The cash flow for each alternative results from the sum, difference, product, or quotient of random variables which influence initial capital investment, operating expenses, revenues, changes in working capital, and other economic factors. Under these circumstances, the measure of economic merit (e.g., the equivalent worth and rate of return values) of the cash flows will also be random variables.

Thus, one can utilize the *PW* as a figure-of-merit in comparing each decision choice. Of course, each decision results in a distribution giving the relative probability for various values of *PW*.

Example 9.1

The heating, ventilating, and air-conditioning (HVAC) system in a commercial building has become unreliable and inefficient. Rental income is being hurt and the annual expenses of the system continue to increase. Your engineering firm has been hired by the owners to

- Perform a technical analysis of the system;
- Develop a preliminary design for rebuilding the system;
- Accomplish an engineering economic analysis to assist the owners in making a decision.

Given the project is completed on schedule, the estimated economic capital investment cost and annual savings in O&M expenses, based on the preliminary design, are shown in the following table. The estimated annual increase in rental income with a modem HVAC system has been developed by the owner's marketing staff and is also provided in the table.

Economic Factor	Estimate
Capital investment	–$521,000
Annual savings	$48,600
Increased annual revenue	$31,000

These estimates are considered reliable because of the extensive information available. The useful life of the rebuilt system, however, is quite uncertain. The estimated probabilities of various useful lives are provided in Table 9.1.

Assume that the MARR = 12% per year and the market value of the rebuilt system at the end of its useful life is zero. Based on this information, what is the *E*(*PW*), the variance *V*(*PW*), and standard deviation *SD*(*PW*) of the project's cash flow? Also, what is the probability of the $PW \geq 0$? What decision would you make regarding the project, and how would you justify your decision using available information?

TABLE 9.1
The Probability Distribution for HVAC System

Useful Life, Years (*N*)	*P*(*N*)
12	0.10
13	0.20
14	0.30
15	0.20
16	0.10
17	0.05
18	0.05

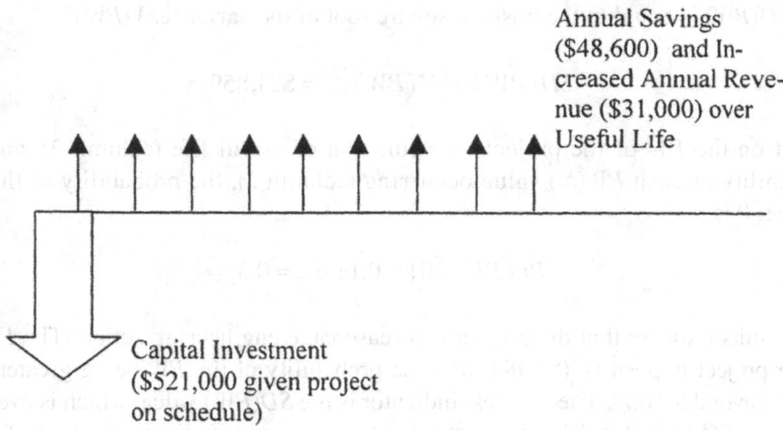

FIGURE 9.3 Cash flow diagram given project on schedule (Example 9.1).

9.2.1 Solution

A cash flow diagram is constructed as shown in Figure 9.3.

According to Equation (9.1), the *PW* of the project's cash flow, as a function of project life (*N*), is

$$PW = -\$521{,}000 + (\$48{,}600 + \$31{,}000)\left[\frac{(1+0.12)^N - 1}{0.12(1+0.12)^N}\right]$$

The calculation of the value of $E(PW)$ = \$9,984 and the value of $E[(PW)2]$ = 5.77527×10^6 (\$)^2 are shown in the following table:

(1) Useful Life (*N*)	(2) *PW*(*N*)	(3) *p*(*N*)	(4) = (2) × (3) *PW*(*N*) × *p*(*N*)	(5) = (2)² [*PW*(*N*)]²	(6) = (3) × (5) *p*(*N*)[*PW*(*N*)]²
12	−\$27,926	0.10	−\$2,793	779.86×10^6	77.986×10^6
13	−\$9,689	0.20	−\$1,938	93.88×10^6	18.776×10^6
14	\$6,605	0.30	\$1,982	43.63×10^6	13.089×10^6
15	\$21,148	0.20	\$4,230	447.24×10^6	89.448×10^6
16	\$34,130	0.10	\$3,413	$1{,}164.86 \times 10^6$	116.486×10^6
17	\$45,720	0.05	\$2,286	$2{,}090.32 \times 10^6$	104.516×10^6
18	\$56,076	0.05	\$2,840	$3{,}144.52 \times 10^6$	157.226×10^6
			E(*PW*) = \$9,984		$E[(PW)^2] = 577.527 \times 10^6 (\$)^2$

The variance of the *PW* is

$$V(PW) = E\left[(PW)^2\right] - \left[E(PW)\right]^2$$
$$= 577.527 \times 10^6 - (9{,}984)^2$$
$$= 477.847 \times 10^6\,(\$)^2$$

The $SD(PW)$ is equal to the positive square root of the variance, $V(PW)$:

$$SD(PW) = \left[V(PW)\right]^{1/2} = \$21,859$$

Based on the PW of the project as a function of useful life (column 2), and the probability of each $PW(N)$ value occurring (column 2), the probability of the PW being ≤ 0 is

$$Pr\{PW \leq 0\} = 0.1 + 0.2 = 0.3$$

The results indicate that the project is a reasonable engineering action. The $E(PW)$ of the project is positive (\$9,984) and the probability of the PW being greater than zero is favorable (0.7). The weakest indicator is the $SD(PW)$ value, which is over two times the $E(PW)$ value. The strategy for risk management is to assure the useful life equal or above 14 years.

The probability that a specific cost, project completion time, or other engineering factor will occur, or that a particular equivalent worth or rate of return value for cash flow will occur, is usually considered to be the long-term relative frequency with which the event (value) occurs or the subjective estimated likelihood that it will occur. Factors such as these, having probabilistic outcomes, are called *random variables.*

The items of information about these random variables that are particularly helpful in decision-making are their expected values and variances, especially for the economic measures of merit of the engineering alternatives. These derived quantities for the random variables are used to make the uncertainty associated with each alternative more explicit, including any probability of loss. Thus, when uncertainty is considered, the variability in the economic measures of merit and the probability of loss associated with the alternatives are both normally used in the decision-making process for risk management.

9.3 INCORPORATING THE IMPACT OF SCHEDULE RISK

For Example 9.1 in Section 9.2, the owner of the commercial building hopes to finish the project within 16 weeks. An engineering firm is consulted to evaluate the schedule risk. Table 9.2 defines the activities within the HVAC project (Example 9.1 in Section 9.2).

In addition to the above information, activity five cannot start until three weeks after the end of activity one. The engineering firm is also consulted on the potential effect upon the overall project completion time (and the critical path) of:

- Cutting the completion time of activity eight "Modify Distribution System" by three weeks?
- Increasing the completion time of activity four "Order Parts for Modified Distribution System" by two weeks?
- Cutting the completion time of activity seven "HVAC Functional Test" by two weeks?

TABLE 9.2
Activities for the HVAC Project

Activity Number	Activity Name	Start Node	End Node	Completion Time (Weeks)
1	Design New HVAC system	1	2	2
2	Modify Distribution System Design	1	3	4
3	Order Parts for Redesigned HVAC	2	4	7
4	Order Parts for Modified Distribution System	3	4	3
5	Design Test Loops	3	5	7
6	Assemble HVAC	4	5	3
7	HVAC Functional Test	5	6	4
8	Modify Distribution System	4	6	6
9	System Integration Test	6	7	2
10	Reliability Acceptance Test	4	7	7

The network diagram is shown below. Note the introduction of node 8 and the two dummy activities 11 and 12 to correctly represent the condition that activity five cannot start until three weeks after the end of activity one.

We then have the following calculation of earliest start times to determine the minimum overall project completion time.

$E_1 = 0$ (by definition)
$E_2 = E_1 + T_{12} = 0 + 2 = 2$
$E_3 = E_1 + T_{13} = 0 + 4 = 4$
$E_4 = \max[E_2 + T_{24}, E_3 + T_{34}] = \max[2 + 7, 4 + 3] = 9$

Now we obviously need to calculate E_8

$E_8 = \max[E_2 + T_{28}, E_3 + T_{38}] = \max[2 + 3, 4 + 0] = 5$
$E_5 = \max[E_4 + T_{45}, E_8 + T_{85}] = \max[9 + 3, 5 + 7] = 12$
$E_6 = \max[E_4 + T_{46}, E_5 + T_{56}] = \max[9 + 6, 12 + 4] = 16$
$E_7 = \max[E_6 + T_{67}, E_4 + T_{47}] = \max[16 + 2, 9 + 7] = 18$

Hence, the minimum overall project completion time is 18 weeks. There is absolutely no possibility to complete the HVAC project with 16 weeks, as hoped by the owner of the commercial building (see Figure 9.4).

To determine the float times and which activities are critical, we need to work out the latest start times.

$L_7 = 18$ (by definition)
$L_6 = L_7 - T_{67} = 18 - 2 = 16$
$L_5 = L_6 - T_{56} = 16 - 4 = 12$
$L_4 = \min[L_7 - T_{47}, L_6 - T_{46}, L_5 - T_{45}] = \min[18–7, 16–6, 12–3] = 9$

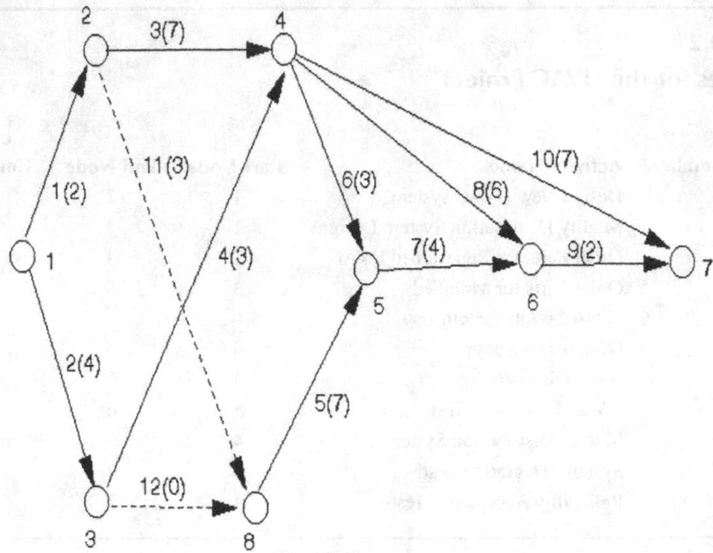

FIGURE 9.4 Network for the HVAC project.

We now need to calculate L_8

$$L_8 = L_5 - T_{85} = 12 - 7 = 5$$
$$L_3 = \min[L_4 - T_{34}, L_8 - T_{38}] = \min[9 - 3, 5 - 0] = 5$$
$$L_2 = \min[L_4 - T_{24}, L_8 - T_{28}] = \min[9 - 7, 5 - 3] = 2$$
$$L_1 = \min[L_2 - T_{12}, L_3 - T_{13}] = \min[2 - 2, 5 - 4] = 0$$

and note that $L_1 = 0$ as required. To calculate the float times, we use the equation $F_{ij} = L_j - E_i - T_{ij}$ to get Table 9.3. Note here that all float times are ≥ 0 as required.

Hence, the critical activities (those with a float of zero) are 1, 3, 5, 6, 7, 9, and 11. This means that there are *two* critical paths, namely 1–11–5–7–9 and 1–3–6–7–9. Thus, the engineering firm finds the following:

- Activity eight "Modify Distribution System" is not critical; therefore, cutting its completion time has no effect upon the overall project completion time or on the critical paths.
- Activity four "Order Parts for Modified Distribution System" has a float time of two weeks, so increasing its completion time by two weeks does not affect the overall project completion time. However, activity four will then become critical, so the critical paths will be affected.
- Activity seven "HVAC Functional Test" is critical, so cutting its completion time by two weeks may reduce the overall project completion time. In fact as activity seven appears in *all* (both) critical paths, we can be sure that the overall project completion time will be reduced by *at least* one time unit (week). The critical paths may, or may not, be affected.

TABLE 9.3
PERT Analysis for the HVAC Project

Activity Number	Activity Name	Start Node	End Node	Latest Start Time	Earliest Start Time	Completion Time (Weeks)	Floating Times
1	Design New HVAC system	1	2	2	0	2	0
2	Modify Distribution System Design	1	3	5	0	4	1
3	Order Parts for Redesigned HVAC	2	4	9	2	7	0
4	Order Parts for Modified Distribution System	3	4	9	4	3	2
5	Design Test Loops	3	5	12	5	7	0
6	Assemble HVAC	4	5	12	9	3	0
7	HVAC Functional Test	5	6	16	12	4	0
8	Modify Distribution System	4	6	16	9	6	1
9	System Integration Test	6	7	18	16	2	0
10	Reliability Acceptance Test	4	7	18	9	7	2
11	Dummy Activity	2	8	5	2	3	0
12	Dummy Activity	3	8	5	4	0	1

Based on the analysis above, the engineering firm proposed to include the impact of schedule risk into the integrated risk management model.

9.4 MONTE CARLO SIMULATION

For Example 9.1, suppose the contracted completion time of the construction project is 80 working days, and the contract stipulates that the penalty because of any delay is

$$D = \$200t^2 \left(\text{instead of } D = \$2{,}000t^2\right)$$

where t is the length of delay (in days). The revised cash flow diagram is shown in Figure 9.5.

Considering the impact of schedule risk, the PW of the project can be calculated as follows:

$$PW = -\$521{,}000 - \$200t^2 + (\$48{,}600 + \$31{,}000)\left[\frac{(1+0.12)^N - 1}{0.12(1+0.12)^N}\right] \quad (9.2)$$

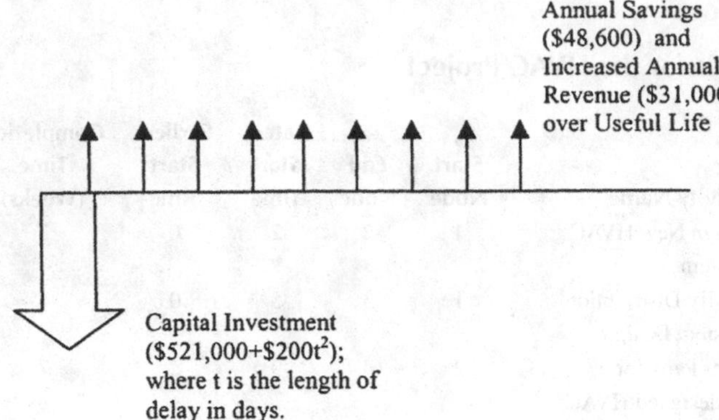

FIGURE 9.5 Revised cash flow diagram including the impact of schedule risk (Example 9.1).

Suppose the delay time, which is driven by the critical path of the project, follows the following extreme value distribution:

$$F_1 = exp\left[-exp\{-0.5(t-10)\}\right] \tag{9.3}$$

The cumulative distribution is plotted in Figure 9.6, which reflects project team's concern with likely project slip:

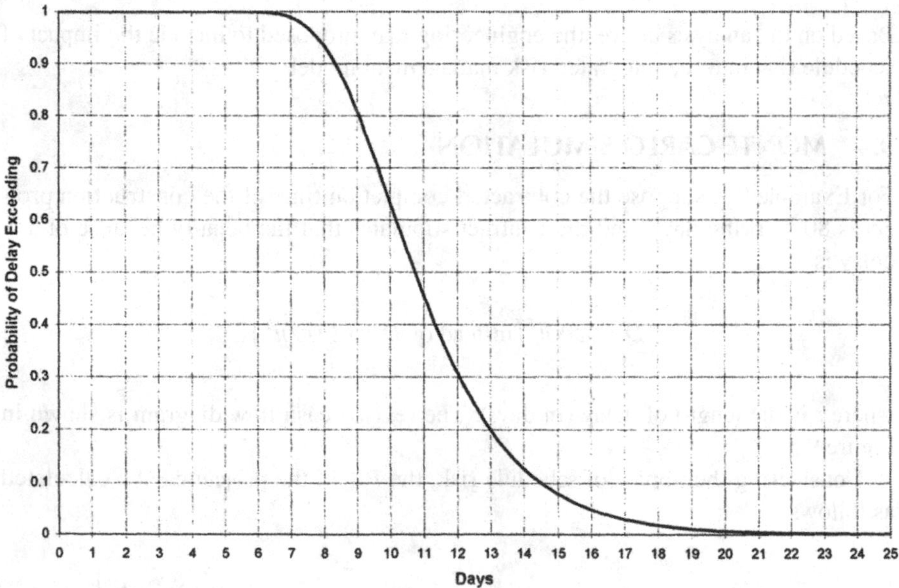

FIGURE 9.6 Cumulative distribution with project delay days.

The problem for integrated risk management now becomes very complicated due to the following factors:

- Nonlinear relationship with *PW* (Equation (9.2));
- Continuous distribution with project delay days;
- Discrete distribution with product's useful life.

Simulation techniques are often used to solve complicated problems like this.

Simulation is the modeling of reality to understand and solve problems. It is the process of replicating the real world based on a set of assumptions and conceived models of reality. Simulation started many years ago with war games but with the advent of computers has spread into many different areas, e.g., aircraft simulation and engineering design. Simulation models can be used for training without the adverse consequences of training with the real-world system that is being modeled (c.f. war gaming or studying an engineering project with potential catastrophic consequences).

For engineering purposes, simulation may be applied to predict or study the performance (like the integrated risk measured by the chance of resulting in a negative PW). With a prescribed set of values for the system parameters (or design variables), the simulation process yields a specific measure of performance or response. As with experimental methods, numerical simulation may be used to obtain (simulated) data, either in lieu of or in addition to actual real-world data. Through repeated simulations, the sensitivity of the system performance to variation in the system parameters may be examined or assessed. In effect, simulation is a method of "numerical or computer experimentation", by which engineers can appraise alternative designs or determine optimal designs.

For very complex problems involving random variables with known (or assumed) probability distributions, Monte Carlo simulation is required. This simulation method enables us to introduce statistical variation into the simulation models using in each simulation a particular set of values of the random variables generated in accordance with the corresponding probability distributions. By repeating the process, a sample of solutions, each corresponding to a different set of values of the random variables, is obtained. A sample from a Monte Carlo simulation is similar to a sample of experimental observations.

For Monte Carlo simulations, we generate parameter values by using a random-number generator (e.g., the digital equivalent of a roulette wheel, which is where the name Monte Carlo comes from). One of the main tasks in a Monte Carlo simulation is the generation of random numbers from prescribed probability distributions; for a given set of generated random numbers, the simulation process is deterministic.

The results of Monte Carlo simulations may be treated statistically; such results may also be presented in the form of histograms, and methods of statistical estimation and inference are applicable. Variance reduction techniques can be used to improve the accuracy of estimates derived from Monte Carlo simulation. The development of computers has resulted in the increased use of Monte Carlo simulation as an important tool for analysis of project risk.

To perform a simulation analysis for integrated risk management, the first step is to construct an analytical model that represents the actual decision situation. This may be as simple as developing an equation for the *PW* of a proposed new HVAC system or as complex as examining the effect of proposed environmental regulations on typical petroleum refinery operations.

The second step is to develop a probability distribution from subjective or historic data for each uncertain factor in the model, such as the system's useful life (*N*) and the project schedule (*t*) for the new HVAC system described above. Sample outcomes are randomly generated by using the probability distribution for each uncertain quantity and are then used to determine a sample of trial outcomes for the model. Repeating this sampling process a large number of times leads to a frequency distribution of trial outcomes for a desired measure of merit, such as the *PW* of the project. The resulting frequency distribution can then be used to make a probabilistic statement about the original problem.

For the probability distribution representing the HVAC system's useful life as shown in Table 9.1, the useful life can be simulated by assigning random numbers to each value such that they are proportional to the respective probabilities. Because two-digit probabilities are given in Table 9.1, a range of random numbers can be assigned to each outcome, as shown in Table 9.5. Next, a single outcome is simulated by choosing a number at random from a table of random numbers, which can be generated for example by the RAND function in Microsoft Excel. For example, if any random number between and including 0.00 and 0.09 is selected, the useful life is 12 years. As a further example, the random number 0.74 corresponds to a life of 15 years.

For the project delay times, which follow an extreme value distribution, a slightly different approach is followed. This approach is generally applicable when we have a cumulative distribution function with a known analytical behavior (see Table 9.4).

Theorem: Let *X* be a random variable with a cumulative distribution function F_x. Let *Y* be a function of *X*, specifically

$$Y = F_x(X)$$

TABLE 9.4
The Probability Distribution for HVAC System

Useful Life, Years (*N*)	*P*(*N*)	Random Number (rounded to two digits beyond the decimal)
12	0.10	0.00–0.09
13	0.20	0.10–0.29
14	0.30	0.30–0.59
15	0.20	0.60–0.79
16	0.10	0.80–0.89
17	0.05	0.90–0.94
18	0.05	0.95–0.99

Then, $Y \sim U(0,1)$. Y is called the integral transform of X.

For the project delay time with probability distribution of Equation (9.3), let

$$Y = F_t = \exp\left[-\exp\{-0.5(t-10)\}\right]$$

Solving for t,

$$t = 10 - 2 \ln(-\ln Y)$$

Using this equation, a sample of $Y \sim U(0,1)$ produces a sample of X from the extreme value distribution for the project delay time.

To reduce the estimation error, it is recommended that a large number of trials be used. This is a formidable task if you were to perform this analysis using hand calculations. A Microsoft Excel Spreadsheet calculation of 100 samples is shown in Table 9.5. In practice, the sample size should be at least 1,000 and usually be above 10,000. Comparing with the results in Section 9.2, the probability of getting negative *PWs* is increased to 0.58 and the average *PWs* = –$2,205.10. This shows that the definite project delay indicated by Figure 9.6 has a significant impact on *PW* of the project due to the contracted penalty for delay.

There are a number of problems related to the use of simulation models:

- Typically, the simulation model has to be run on the computer for a considerable time in order for the results to be statistically significant – hence, simulations are expensive (take a long time) in terms of computer time.
- Results from simulation models tend to be highly correlated, meaning that estimates derived from such models can be misleading (correlation is a statistical term meaning that two (or more) variables are related to each other in some way).
- In the event that we are modeling an existing system, it can be difficult to validate the model (computer program) to ensure that it is a correct representation of the existing system.
- If the simulation model is very complex, then it is difficult to isolate and understand what is going on in the model and deduce cause and effect relationships.
- Integrated risk management models simulate the operations of an engineering project and require large amounts of data relating to quality, quantity, schedule, prices, manpower, overheads, materials, etc. The advantage of integrated risk management models is that random variations can be modeled to provide a range for, say, the possible profit and risk from a project.

At the heart of Monte Carlo simulation is the generation of random numbers. Most spreadsheet packages includes a RAND () function that returns a random number between zero and one. Other advanced statistical functions, such as NORMSINV (), will return the inverse of a cumulative distribution function (the standard normal distribution in this case). This function can be used to generate random normal

TABLE 9.5
Monte Carlo Simulation of the HVAC Project

Index	Random Number −1(0,1)	Useful Life	*PW* of Annual Saving and Revenue	Random Number −2(0,1)	Project Delay (days)	Present Worth of Project	Counter (1 denotes Negative *PW*)
1	0.37	14	$527,602.19	0.295811	9.60552	($2,624.41)	1
2	0.89	16	$555,129.30	0.771885	12.70247	$17,994.02	0
3	0.29	13	$511,314.45	0.45933	10.50209	($20,714.94)	1
4	0.09	12	$493,072.19	0.584551	11.24385	($40,570.22)	1
5	0.63	15	$542,144.81	0.119073	8.48962	$13,937.45	0
6	0.47	14	$527,602.19	0.671445	11.84098	($7,418.69)	1
7	0.10	13	$511,314.45	0.110263	8.418645	($16,772.90)	1
8	0.05	12	$493,072.19	0.68861	11.97192	($42,260.51)	1
9	0.74	15	$542,144.81	0.906784	14.64863	($313.42)	1
10	0.82	16	$555,129.30	0.6111	11.41654	$21,095.56	0
11	0.95	18	$577,073.74	0.051903	7.830718	$49,941.72	0
12	0.47	14	$527,602.19	0.019881	7.26885	$1,318.57	0
13	0.40	14	$527,602.19	0.873454	14.00053	($12,999.28)	1
14	0.57	14	$527,602.19	0.471679	10.57148	($4,573.43)	1
15	0.45	14	$527,602.19	0.528382	10.89904	($5,276.71)	1
16	0.98	18	$577,073.74	0.714364	12.17913	$41,240.61	0
17	0.89	16	$555,129.30	0.02858	7.463261	$28,559.27	0
18	0.40	14	$527,602.19	0.25203	9.358436	($2,155.84)	1
19	0.86	16	$555,129.30	0.253299	9.365736	$25,357.60	0
20	0.99	18	$577,073.74	0.832886	13.39809	$38,122.85	0
21	0.21	13	$511,314.45	0.477649	10.60524	($20,932.67)	1
22	0.83	16	$555,129.30	0.316731	9.720995	$24,679.52	0
23	0.39	14	$527,602.19	0.654228	11.71463	($7,121.07)	1
24	0.07	12	$493,072.19	0.979214	17.72594	($59,348.71)	1
25	0.31	14	$527,602.19	0.763273	12.61763	($9,318.27)	1
26	0.41	14	$527,602.19	0.328195	9.78382	($2,970.12)	1
27	0.88	16	$555,129.30	0.109527	8.412581	$27,052.15	0
28	0.47	14	$527,602.19	0.200515	9.05143	($1,590.65)	1
29	0.08	12	$493,072.19	0.269307	9.457044	($36,871.38)	1
30	0.17	13	$511,314.45	0.137354	8.628569	($17,130.77)	1
31	0.42	14	$527,602.19	0.199085	9.042539	($1,574.56)	1
32	0.83	16	$555,129.30	0.668834	11.82151	$20,154.48	0
33	0.49	14	$527,602.19	0.689979	11.9826	($7,756.07)	1
34	0.97	18	$577,073.74	0.763747	12.62224	$40,141.65	0
35	0.01	12	$493,072.19	0.323979	9.760746	($37,455.03)	1
36	0.29	13	$511,314.45	0.960196	16.40709	($36,604.80)	1
37	0.55	14	$527,602.19	0.374334	10.03509	($3,468.12)	1
38	0.71	15	$542,144.81	0.382089	10.07727	$10,989.67	0
39	0.11	13	$511,314.45	0.470919	10.5672	($20,852.11)	1
40	0.86	16	$555,129.30	0.123984	8.527967	$26,856.68	0

(Continued)

TABLE 9.5 ***(Continued)***
Monte Carlo Simulation of the HVAC Project

Index	Random Number −1(0,1)	Useful Life	*PW* of Annual Saving and Revenue	Random Number −2(0,1)	Project Delay (days)	Present Worth of Project	Counter (1 denotes Negative *PW*)
41	0.05	12	$493,072.19	0.437719	10.38189	($38,706.17)	1
42	0.78	15	$542,144.81	0.434739	10.36542	$10,400.62	0
43	0.86	16	$555,129.30	0.128057	8.559178	$26,803.35	0
44	0.35	14	$527,602.19	0.147351	8.700631	($967.91)	1
45	0.31	14	$527,602.19	0.457703	10.49299	($4,408.09)	1
46	0.65	15	$542,144.81	0.021835	7.317277	$15,790.56	0
47	0.47	14	$527,602.19	0.724392	12.26379	($8,437.86)	1
48	0.93	17	$566,722.59	0.318167	9.728878	$36,257.48	0
49	0.93	17	$566,722.59	0.799686	12.99637	$28,832.03	0
50	0.42	14	$527,602.19	0.865106	13.86338	($12,617.14)	1
51	0.71	15	$542,144.81	0.642246	11.62935	$7,620.63	0
52	0.91	17	$566,722.59	0.823409	13.27668	$28,095.55	0
53	0.41	14	$527,602.19	0.121902	8.511808	($642.90)	1
54	0.83	16	$555,129.30	0.677794	11.88881	$19,994.92	0
55	0.17	13	$511,314.45	0.087767	8.221692	($16,445.17)	1
56	0.53	14	$527,602.19	0.010011	6.946133	$1,777.32	0
57	0.95	18	$577,073.74	0.732994	12.33839	$40,850.16	0
58	0.33	14	$527,602.19	0.771013	12.69376	($9,510.97)	1
59	0.59	14	$527,602.19	0.848425	13.61123	($11,924.35)	1
60	0.22	13	$511,314.45	0.10722	8.393422	($16,730.50)	1
61	0.49	14	$527,602.19	0.138507	8.637009	($857.60)	1
62	0.88	16	$555,129.30	0.873197	13.99618	$14,539.99	0
63	0.80	16	$555,129.30	0.115425	8.460587	$26,971.14	0
64	0.84	16	$555,129.30	0.180392	8.923946	$26,165.62	0
65	0.56	14	$527,602.19	0.806165	13.06989	($10,480.02)	1
66	0.72	15	$542,144.81	0.55305	11.04746	$8,940.17	0
67	0.88	16	$555,129.30	0.104436	8.369996	$27,123.62	0
68	0.91	17	$566,722.59	0.210335	9.111842	$37,420.02	0
69	0.77	15	$542,144.81	0.707928	12.12603	$6,440.76	0
70	0.32	14	$527,602.19	0.278191	9.507142	($2,436.38)	1
71	0.38	14	$527,602.19	0.256916	9.386492	($2,208.43)	1
72	0.65	15	$542,144.81	0.691624	11.99547	$6,755.68	0
73	0.18	13	$511,314.45	0.024435	7.376992	($15,127.55)	1
74	0.81	16	$555,129.30	0.968801	16.90315	$5,557.66	0
75	0.27	13	$511,314.45	0.233791	9.252287	($18,246.03)	1
76	0.36	14	$527,602.19	0.515746	10.82456	($5,114.91)	1
77	0.27	13	$511,314.45	0.676108	11.87604	($23,789.58)	1
78	0.36	14	$527,602.19	0.517274	10.83351	($5,134.30)	1
79	0.32	14	$527,602.19	0.782009	12.80575	($9,796.54)	1
80	0.33	14	$527,602.19	0.941159	15.60548	($17,750.91)	1

(Continued)

TABLE 9.5 *(Continued)*
Monte Carlo Simulation of the HVAC Project

Index	Random Number −1(0,1)	Useful Life	*PW* of Annual Saving and Revenue	Random Number −2(0,1)	Project Delay (days)	Present Worth of Project	Counter (1 denotes Negative *PW*)
81	0.30	14	$527,602.19	0.070368	8.047856	$125.39	0
82	0.07	12	$493,072.19	0.948463	15.87824	($53,139.65)	1
83	0.09	12	$493,072.19	0.287739	9.5606	($37,068.32)	1
84	0.97	18	$577,073.74	0.457464	10.49166	$45,066.25	0
85	0.14	13	$511,314.45	0.688415	11.97041	($24,014.61)	1
86	0.10	13	$511,314.45	0.898777	14.47508	($30,638.33)	1
87	0.01	12	$493,072.19	0.052214	7.834764	($34,066.16)	1
88	0.14	13	$511,314.45	0.821905	13.25795	($27,262.87)	1
89	0.42	14	$527,602.19	0.637431	11.59564	($6,843.69)	1
90	0.00	12	$493,072.19	0.108114	8.400876	($34,985.28)	1
91	0.67	15	$542,144.81	0.736224	12.3669	$5,850.79	0
92	0.79	15	$542,144.81	0.111249	8.426739	$14,043.82	0
93	0.08	12	$493,072.19	0.691872	11.99742	($42,321.62)	1
94	0.30	14	$527,602.19	0.987282	18.71661	($28,428.97)	1
95	0.82	16	$555,129.30	0.353896	9.92396	$24,280.80	0
96	0.63	15	$542,144.81	0.354385	9.926621	$11,291.03	0
97	0.60	15	$542,144.81	0.080073	8.14766	$14,506.38	0
98	0.31	14	$527,602.19	0.332766	9.808806	($3,019.08)	1
99	0.25	13	$511,314.45	0.456271	10.48499	($20,679.05)	1
100	0.88	16	$555,129.30	0.109935	8.415948	$27,046.48	0
						Average *PW* = ($2,205.10)	Total number of negative *PWs* = 58

Probability of Resulting in Negative *PWs* = 58/100 = 0.58

deviates. Here, the simulated outcome is based on the mean and standard deviation of the probability distribution and on the random normal deviate, which is a random number of standard deviations above or below the mean of a standardized normal distribution. For normally distributed random variables, the simulated outcome is based on the following equation:

$$\text{Outcome value} = \text{mean} + [\text{random normal deviate} \times \text{standard deviation}] \quad (9.4)$$

Readers may consider Example 9.1 while including the uncertainty with annual savings due to the cost risk with O&M expenses. The annual savings follow a normal distribution (see Figure 9.7 for a revised cash flow diagram).

After a simulation model has been put onto the computer, it is relatively easy to "play" with it to examine the effect of changes (e.g., in raw material costs) on the

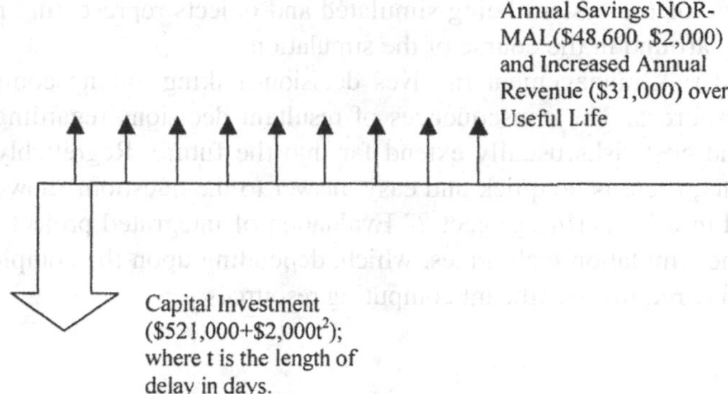

FIGURE 9.7 Revised cash flow diagram including the uncertainty with annual savings (Example 9.1).

engineering project. Computer packages are available to help companies construct risk management models:

- bottom-up models which consider every individual element of project data; and
- top-down models which work from the project master control sheet.

Once we have the model, we can use it to:

- Understand the current system, typically to explain why it is behaving as it is. For example, if we are experiencing long delays in production in a factory, then why is that occurring – what factors are contributing to these delays?
- Explore extensions (changes) to the current system, typically to try and improve it. For example if we are trying to increase the output from a factory, we could:
  - add more machines; or
  - use machines more hours per day or more days per week;
  - speed up existing machines; or
  - reduce machine idle time by better maintenance.
  - Which of these factors (or combination of factors) will be the best choice to increase output. Note here that any change might reduce congestion at one point only to increase it at another point, so we must bear this in mind when investigating any proposed changes.
- Design a new system from scratch, typically to try and design a system that fulfills certain (often statistical) requirements at minimum cost. For example, in the design of an airport passenger terminal, what resource levels (customs, seats, baggage facilities, etc.) do we need and how should they be sited in relation to one another.

A comparatively recent development in simulation for integrated risk management are packages which run on PCs with *animation* – essentially on screen one sees a representation of the system being simulated and objects representing risk parameters moving around in the course of the simulation.

Integrated risk management involves decision-making among competing uses of scarce resources. The consequences of resultant decisions regarding technical, schedule, and cost risks usually extend far into the future. Regrettably, as shown in this chapter, there is no quick and easy answer to the question, "how should risk be managed in engineering projects?" Evaluation of integrated project risk is possible with the simulation techniques, which, depending upon the complexity of the modeling, may require significant computing resources.

9.5 DIGITAL TRANSFORMATION IN THE FACE OF COVID-19

The modification of data characteristics for better access or storage is known as data transformation. Data's format, structure, or values may all undergo transformation. In data analytics, transformation often takes place after data has been extracted or loaded. The process of changing data from one format or structure to another is known as data transformation in computing. It is a crucial component of the majority of data management and integration jobs, including application integration, data wrangling, data warehousing, and data integration.

The COVID-19 epidemic earlier this year caused a large portion of the world to migrate online, speeding a long-term digital revolution. Many people began working from home, children with at-home Internet connection started attending classes online, and numerous businesses adapted digital business models to continue operating and keep certain revenue streams. In the meantime, researchers used artificial intelligence (AI) to learn more about the virus and hasten the quest for a vaccine. Mobile applications were also created to aid "monitor and trace" the spread of the pandemic. Shortly after the epidemic, Internet traffic in some countries rose by as much as 60%, highlighting the digital acceleration that the pandemic sparked.

While these actions highlight the digital transformation's enormous promise, the pandemic has also highlighted the holes that still need to be filled. Despite the fact that certain digital barriers have closed quickly in recent years, others have not, leaving some people behind in the COVID-driven digital acceleration. Concerns about privacy and digital security have also become more urgent as a result of our greater reliance on digital solutions.

This poses a significant problem for nations. The "pre-COVID" structures of economies and society are unlikely to return; the crisis has eloquently illustrated the potential of digital technologies, and certain changes may already be irreversible. In light of the potential for greater reliance on digital technologies in the future for jobs, education, health, government services, and even social interactions, failing to ensure widespread and reliable digital access and efficient use runs the risk of escalating inequality and could obstruct nations' efforts to recover more quickly from the pandemic.

The expanding significance of digital technology and communications infrastructures in our day-to-day lives has been brought to light, and it shows how governments are increasingly putting digital initiatives at the forefront of their policy agendas. Now is the time to achieve an inclusive digital transformation, with coordinated and comprehensive plans that create resilience and bridge digital gaps for a post-COVID age, as nations struggle to respond to and recover from the COVID-19 crisis.

9.5.1 Risk Engineering of Digital Transformation: COVID-19 Has Upped the Bar

Fast and dependable connectivity enables the usage of linked devices in crucial contexts, such as health, manufacturing, and transportation, and makes interactions between people, organizations, and machines easier. Over time, connectivity has constantly increased.

The COVID-19 has increased the stakes around digital access and participation as a result of the surge in economically and socially facilitated by digital activity, highlighting the reality that connectivity and the use of digital technology are dynamic goals. However, in areas where the pandemic has served as a stimulant, such as telework, e-commerce, e-health, and e-payments, it is likely that online activity will continue to be high. Some online activity may fall if COVID-19 cures start to emerge and enable more in-person encounters. This keeps pressure on creating high-quality connectivity and increases people's and businesses' capacity to adopt increasingly complex digital solutions.

As governments adjust their strategies in response, they should keep in mind that increased reliance on digitalization could risk opening new digital divides and/or exacerbating those that have proved persistent over the years. For example, Internet users ranged from over 95% to less than 70% of the adult population in 2019, and there are important demographic differences in Internet use. Although 58% of those aged 50–74 used the Internet daily in 2019 – up from only 30% in 2010 – this remains well below the average share of daily Internet users aged 16–24 (the so-called "digital natives"), which was close to 95%. People with greater skill or wealth levels use the Internet and online activities better and are better able to access knowledge, job opportunities, and health and education services, but there are still persisting skills and inequalities among demographic groups and countries. Another crucial policy objective is closing the gender gap in digital technology. Women are more likely to report workplace stress related to regular computer use at work, while men are more likely to demonstrate high-demand abilities in these industries.

The dispersion and use of digital technology within enterprises also differ significantly. Prior to the pandemic, e-commerce represented 19% of company revenue (mostly from business-to-business transactions), despite major differences between large and small businesses (24% vs. 9%). And although the usage of big data has grown over time, it continues to vary greatly between nations and industries. Big data was utilized by over 25% of information and communication technology companies in the European Union in 2018, compared to just 10% of all companies.

9.5.2 Digital Revolution: A Top Priority for Governments' Policy Objectives

Prior to the COVID-19 outbreak, many governments had increased their strategic approach to the digital transition. Additionally, governments are paying more attention to new digital technologies like AI, blockchain, and 5G infrastructure, which is essential for supporting improved mobile broadband, Internet of Things (IoT) gadgets, and AI applications. In the previous three years, several nations, including Australia, Austria, Colombia, France, Germany, Korea, Spain, the United Kingdom, and the United States, have released national 5G strategies. By the middle of 2020, 60 nations will have national AI strategies in place. Additionally, policymakers are paying more attention to blockchain technology and quantum computing. Australia, the People's Republic of China, Germany, India, and Switzerland are among the nations that have published a blockchain strategy, while France and Italy are creating one.

New business models and marketplaces are fundamentally driven by the positive feedback loop between digital innovation and transformation, and digital technologies have the ability to boost the scientific and research systems that are so crucial to the COVID-19 response and recovery of nations. However, nations are also becoming aware that how these technologies are employed can endanger human-centered values, as well as consumer protection, privacy, and security. This provides them more motivation to define strategic goals, particularly on a global scale, where the AI Principles are only one example of like-minded nations working together to develop reliable technologies.

9.5.3 Need to Create a Digital Future that Is More Inclusive

Although hopeful, this strategic trend might not be sufficient to guarantee a robust and more inclusive digital future. A coordinated, whole-of-government policy approach to digital transformation is required in light of the COVID-19 dilemma. The right policy environment will depend on a delicate balancing act that will differ for every country due to cultural, social, and economic considerations.

A solution is provided by risk engineering. The framework, which is based on seven building blocks including access, use, innovation, trust, jobs, society, and market openness, assembles the regulations that governments must take into account in order to create a shared digital future that enhances people's lives and fosters economic growth and well-being. In light of the COVID-19 crisis, the importance of these pillars, as well as the metrics and policy recommendations that support them, has increased.

1. Access: The COVID-19 crisis has highlighted the significance of communications infrastructures and services, as well as access to and effective administration of data, by driving many businesses and schools online and enacting lockdowns and social distancing measures. Upgrading networks to the next evolution of fixed and wireless broadband, improving access to and the sharing of data, and addressing rural/urban inequalities in broadband

access and underserved socioeconomic groups can all contribute to economic and social advantages.

2. Use: Governments must endeavor to guarantee that all workers have the skills necessary to succeed in the digital economy as more people and businesses "go digital" in the wake of the COVID-19 issue. They also need to do more to promote adoption among small- and medium-sized businesses (SMEs). People who have a diverse skill set in a technologically advanced environment, including reading, numeracy, and problem solving, can be anticipated to use digital tools more effectively, engage in more complex online activities, and adapt to changes in the digital landscape more successfully.
3. Innovation: Digital innovation is a key factor in the digital transformation since it gives rise to new products and services, opens up new markets and business models, and can increase efficiency in the public sector and other areas. A strong reaction to and recovery from the crisis can be supported by encouraging entrepreneurship, enabling greater digital transformation of scientific research, and encouraging investment in research and development.
4. Trust: Following COVID-19, there has been a higher reliance on digital technologies; therefore, it is important to pay more attention to building confidence in the digital environment, particularly in terms of digital security, but also in terms of privacy, data protection, and consumer protection. As the epidemic spread, phishing and frauds involving coronaviruses increased as con artists profited from the widespread shift to Internet activity. The majority of nations have implemented whole-of-government digital security policies, but frequently these strategies are not connected with the overall national digital objectives and lack an autonomous budget, evaluation tools, and metrics.
5. Jobs: Organizations and markets are already changing as a result of the digital transformation, which has led to significant uncertainty about the future of employment. The epidemic has caused a rise in teleworking across many businesses and cast doubt on the future of some employment, making the situation even more murky. Policymakers will need to reexamine labor market institutions and laws as they deal with the economic effects of the crisis and as automation spreads throughout economies. They must also endeavor to guarantee that workers who are displaced are not left behind.
6. Society: Extra care is required to support people's well-being as they spend more time online during the epidemic, whether for job, school, or social engagement. Governments ought to take use of this chance to address the wide spectrum of social problems that the digital transformation has brought to light, including concerns like data-driven healthcare, false information, and scree addiction, among many others.
7. Market openness: Concerns about market consolidation have been raised by the COVID-19 dilemma as start-ups and SMEs struggle to survive and as major technology giants increasingly control our digital life. Governments must take into account the effects on economic dynamics and inclusiveness as fewer businesses mediate access to the online world.

The way we live and work will undoubtedly continue to change as a result of digital technology, regardless of how the crisis and its aftermath play out. The IoT and 5G's advent will accelerate the generation of data, making the existing policy debates over data governance, privacy, and security even more urgent. The relevance of data transfers between companies may increase as businesses balance the advantages and disadvantages of increasing automation, particularly in manufacturing plants, to improve resilience against future health crises.

Governments will encounter complicated, interconnected concerns when they review their current digital policies in light of the COVID-19 dilemma, which calls for concerted international coordination, cooperation, and dialogue. Companies will continue to actively cooperate with nations to assist their digital transformation and help them navigate the post-pandemic future through evidence-based analysis, policy advice, and the establishment of international standards in areas like AI and privacy.

9.5.4 Managing Big Data Transformation's Complexity, A Driver of Risk

Businesses must provide a dependable, high-value experience if they don't want to lose clients to competitors who can. And they are seeking assistance from big data technologies. With big data, they can get to know their customers better, learn their habits, and anticipate their needs to ultimately deliver a better customer experience.

Data transformation is a method for converting and translating data across different formats. The format, complexity, structure, and volume of the data all influence the tools and techniques used for data transformation.

Regardless of the format used to represent data, it enables a developer to convert between XML, non-XML, and Java data formats enabling quick integration of heterogeneous applications. The top eight data transformation techniques are shown here in alphabetical order.

9.5.4.1 Aggregation

The process of gathering raw data and expressing it in a condensed form for statistical analysis is known as data aggregation. For example, aggregating raw data over a specified time period can yield statistics like average, minimum, maximum, sum, and count. You can analyze the aggregated data to learn more about specific resources or resource groupings after the data has been compiled and written up as a report. Time aggregation and geographical aggregation are the two categories of data aggregation.

9.5.4.2 Attribute Construction

This technique aids in streamlining the data mining process. To assist the mining process, additional attributes are generated and added from the provided set of attributes in attribute construction or feature construction.

9.5.4.3 Discretization

Data discretization is the process of assigning a particular data value to each interval and transforming continuous data attribute values into a finite collection of intervals. There are many different discretization techniques, ranging from simple ones like equal-width and equal-frequency to far more complex ones like Minimum Description Length (MDLP).

9.5.4.4 Generalization

Data generalization is a technique for creating progressively more layers of summary data in an evaluational database to obtain a more thorough understanding of a situation or problem. Online Analytical Processing can benefit from data generalization (OLAP). The basic purpose of OLAP is to quickly respond to multidimensional analytical queries. The technique is useful for implementing online transaction processing (OLTP). Data entry and retrieval transaction processing applications, in particular, are referred to as "transaction-oriented applications", or "OLTP", and are managed and facilitated by a class of systems.

9.5.4.5 Integration

Data pre-processing entails merging data from several sources and giving users a single picture of those data. This process is known as data integration. It functions by combining the data from many data sources and includes several databases, data cubes, or flat files. For data integration, there are primarily two major approaches: the tight coupling strategy and the loose coupling approach.

9.5.4.6 Manipulation

Data manipulation is the process of modifying data to improve its readability and organization. Tools for data manipulation aid in finding patterns in the data and transforming it into a format that can be used to provide insights into things like financial data, consumer behavior, etc.

9.5.4.7 Normalization

Data normalization is a technique to change the format of the source data for efficient processing. Data normalization's main goal is to reduce or even eliminate redundant data. It has many benefits, like improving the efficiency of data mining algorithms and speeding up data extraction, among others.

9.5.4.8 Smoothing

A method for finding trends in noisy data while the trend's form is unclear is data smoothing. The approach can be used to spot patterns in the stock market, the economy, consumer attitude, etc.

9.5.5 Conclusion

For the greatest outcomes, information analysis needs structured and easily accessible data. Organizations can change the format and structure of raw data through data transformation as needed. Based on risk engineering and management, data transformation helps engineers to transform data to perform analytics, the process of discovering, interpreting, and communicating significant patterns in data, effectively.

9.6 APPLY WANG ENTROPY TO ANALYZE MODE CONFUSION, A CHALLENGE TO RISK MANAGEMENT AT THE AGE OF AUTONOMY

Mode confusion means a condition in which drivers or pilots become confused about the status of the system and interact with it in an incorrect manner.

- uncertainty about an automation system's present mode of operation might cause the information to be interpreted incorrectly or to be used inappropriately.

9.6.1 CASE STUDY 1: DEVELOPING A CLEAR INTERFACE FOR CONTROL TRANSFER IN A LEVEL 2 AUTOMATED DRIVING SYSTEM

9.6.1.1 How Can We Eliminate Mode Confusion by Intuitively Managing the Transition between Autonomous and Manual Driving Modes?

Mode confusion has affected you if you've ever accidentally pressed the accelerator while your car was in park or tried to type into the wrong window on your computer screen. Mode confusion is reported to be a factor in 50% of airplane crashes, and it's much more dangerous in cars. Drivers aren't typically highly skilled experts with thousands of miles and countless minutes to respond to issues; they also deal with dozens of issues every minute. AV drivers must be completely aware of whether they are driving manually or in autonomous mode. For AVs to be secure, all confusion in this context needs to be completely removed.

9.6.1.2 Developing a Clear Interface for Control Transfer in a Level 2 Automated Driving System

Designing an efficient HMI for L2 vehicles that takes into account both driving with automation enabled and the transition where drivers take back control from the system has received a lot of attention. There are three issues that need to be resolved when building an interface for L2 vehicles:

- how to avoid mode confusion when presenting information about the system's status;
- how to communicate take-over requests to drivers; and
- how to persuade drivers to focus once again on the road. A feedback system could be helpful in correcting these issues. Three different sources of feedback – visual, aural, and tactile – can be used singly or in combination.

Prior research that indicated L2 vehicles has been addressed.

- The first issue was mode confusion, which was resolved by giving a helpful presentation of system status through an LED icon (car between lanes).
- The second issue was the delivery of take-back control requests; this was resolved by redundant audio and visual feedback to guarantee the best take-back control responses when necessary.

- The third issue was to get drivers to pay attention again to hazardous situations. This was accomplished by giving participants feedback on the road geometry and objects detected (along with LED icons and beeps), which encouraged them to be attentive and turn their focus to the hazards up ahead.

9.6.1.3 What Is Mode Awareness and Automated Driving, and How Is It Quantified?

9.6.1.3.1 Mode Confusion and Mode Errors

Deficient mode awareness may result from mode confusion, among other things. It can be characterized as a form of automation surprise in which the system behaves differently than what the user would have anticipated. When a user suffers mode confusion, they become unaware of which system is currently in use or the required behavior for each mode. Because it can result in mode errors, mode confusion poses a serious significant risk. This term describes behavior that fits the assumed but not the actual active automation level. It comes about as a result of the mental model combining incorrect information. When a driver switches between two or more vehicles, or when different systems are accessible in one vehicle, mode confusion may occur. The latter situation has a greater risk of mode confusion. If the systems seem similar to the user, as they do in the cases of partially automated driving (PAD) and conditionally automated vehicles (CAD), the likelihood of mode confusion grows even more. As a result, when operating a vehicle in PAD, drivers may take part in non-driving-related tasks (NDRTs) and omit to do their monitoring task. If the system hits its limit without the driver realizing it, this can be extremely hazardous and cause crashes.

When the system hits a system limit, mode confusion, for instance, could become obvious. Additionally, a person may grip the driving wheel while performing CAD, continuously push unrelated keys, or display confused facial expressions. However, these behaviors may differ between participants and may not always be present during a drive, which reduces the comparability of these behaviors. Further, without a follow-up interview, it is impossible to say with certainty if this behavior is due to mode awareness. For instance, a user might place their hands on the steering wheel out of habit or comfort rather than mistaking the automated mode that is currently in effect.

The participants' understanding of Human Machine Interaction (HMI) and how to handle the systems' (de-)activation is tested. That primarily serves as an indicator for mode confusion in the drive since uncertainty about the associated icons for each mode and the proper button for activating and deactivating the systems can easily result in confusion regarding the currently active system.

Additionally, it is highly beneficial to include a variety of questions, such as those pertaining to how the events during the silent system error were perceived, participation in NDRTs during PAD, absence of participation in NDRTs during CAD, automation surprises, and other subjective data.

9.6.1.3.2 We Need to Address Mode Confusion Right Away

In a safety-critical situation, a human driver who is aware, engaged, and unimpaired performs far better than a machine driver. Road safety is more challenging when mode confusion is present, along with the instantaneous doubt it causes over who is responsible for doing the driving activity.

In summary, a driver's confusion about the dynamic operating modes of an Automated Vehicle (AV), and thereby their confusion about their driving responsibilities can compromise safety.

9.6.1.3.3 Safety Implications of Autonomous Driving Assist Alerting Variability

Mode confusion, which has been documented in cars with limited autonomy, is made possible by complacency. When a driver believes the car is in one mode when it is actually in another, mode confusion develops. In the event of unintentional autopilot disengagements, already complacent drivers may believe the vehicle is capable of handling the driving task, so they immediately engage in a distracting task after briefly removing their hands from the wheel, unaware that they, and not the autopilot system, are in charge of the driving task. Mode confusion may result in no one driving the vehicle, which has resulted in fatalities.

The unexplained autopilot shutoffs in tests raises the possibility that subsequent complacency and mode confusion could result in actual transportation safety issues. This is supported by the high consistency of alerts in the highway tests. In vehicles with L2+ systems, complacency is becoming a bigger issue as drivers form poor monitoring habits and lengthen their periods of distraction. Unexpected autopilot disengagements only occurred in a small percentage of cases (3.4%), but inattentive drivers can easily overlook a situation like this if there isn't a clear warning.

9.6.1.4 Conclusion

A driver's confusion about the dynamic operating modes of an AV, and thereby their confusion about their driving responsibilities can compromise safety.

9.6.2 Case Study 2: Pilots of 777s and 787s Warned over Pitch-Guidance Mode Slip before Take-Off

The U.S. safety authorities have warned operators of the Boeing 777 and 787 about a potential mode confusion during take-off that might cause the aircraft to depart with the incorrect pitch-control guidance. There have been instances where the autopilot flight-director system's "altitude hold" mode was mistakenly engaged prior to take-off.

The outcome, according to the U.S. FAA, was extremely low initial climb rates in one case and "don't sink" alerts from the aircraft's ground-proximity warning system in another. However, the U.S. FAA has not named the specific operators involved. According to an analysis of the occurrences, the incorrect pitch mode had been turned on "without the crew's recognition" prior to take-off.

The crews activated the take-off/go-around switch after taking off to reestablish appropriate pitch guidance. When the flight director is initially turned on, the default roll and pitch mode, according to the FAA, is "take-off/go-around".

However, on both the 777 and 787, there are several ways in which this pitch mode can accidentally be changed to "altitude hold". Inadvertently selecting the "altitude hold" switch on the mode control panel is one of them, as is selecting the "flight level change" or "vertical speed" switches when the selected altitude is less than 20 feet from the barometric altitude. The FAA has also drawn attention to other specific instances that can result in the transition, which are unique to each type of aircraft and can happen, for example, while re-aligning the air data inertial reference systems on a 777.

Pressing the take-off/go-around switch will have no impact on the pitch mode or guidance if the aircraft is in "altitude hold" mode when on the ground. If the plane takes off in this mode and the displayed barometric altitude is close to the selected altitude on the mode control panel, the flight director will direct a nose-down pitch shortly after the plane takes off. Although the Boeing flight crew operations manual does explain "altitude hold" latching and "take-off/go-around" behaviors, this particular scenario is not specifically covered.

It is apparent that this specific system behavior may not be known to all pilots of these aircraft. On March 4, Boeing issued an operator message regarding the problem with instructions on how to return the aircraft to the proper "take-off/go-around" mode while still on the ground.

The FAA advises twinjet model operators to notify crews to the problem and implement the suggested solutions as soon as possible. It continues by stating that the crew must maintain situational awareness to make sure the aircraft is in the appropriate mode at all times during any given phase of flight.

Analysis of the "altitude hold" and "take-off/go-around" logic in other Boeing aircraft, such as the 757, 767, 747-400, and 747-8, is currently under way. According to the FAA, no incidents of this nature have been documented thus far.

9.6.3 Apply Wang Entropy to Analyze Mode Confusion

Automation systems unavoidably become more sophisticated as they gain the capacity to carry out more tasks and make use of more input sources. The number of automation modes that a user must comprehend tends to grow as complexity grows. The capacity of automation to help the user with individual demands in a variety of settings may be improved by the increased number of modes. The user might perceive the information being provided incorrectly or submit input that is improper for the present mode, however, if the automation user is ignorant of the current system mode. Mode confusion is presenting a challenge to integrated risk management at the age of autonomy.

Regarding mode confusion, the procedure for developing Safety Mechanisms for identifying onboard diagnosis using Fault Tree Analysis (FTA) is summarized as follows:

1. For mode confusion, develop quantitative FTA/calculate importance measures
   - Generate Minimal Cut Sets (MCSs): reflecting the collections of faults/failures with Signals for Modes Indications;
   - Generate cut set importance to prioritize the collections of faults/failures with Signals for Modes Indications;
   - Generate Fussell-Vesely importance (for basic events): to prioritize the faults/failures with individual signals for Modes Indications.
2. For mode confusion, develop Safety Mechanisms for onboard Diagnosis of Mode Confusion: at every stage of a fault diagnosis, the individual signal

(or fault/failure of the individual signal) whose Fussell-Vesely importance is nearest to 0.5 is chosen to be inspected/diagnosed;

- This approach will lead to identifying the cause of mode confusion with a minimum number of average inspections (diagnoses); thus, we can develop Safety Mechanisms to prevent/mitigate.
- As presented in Chapter 1, the average number of inspections (diagnoses) is lower-bounded by Wang Entropy.

Here, Wang Entropy, the entropy of cut set importance, measures information uncertainty for diagnosing the mode confusion.

BIBLIOGRAPHY

Anon. (1989), *Risk Management Concepts and Guidance*, Defense Systems Management College, Fort Belvoir VA.

Batson, R. G. (1987), "Critical Path Acceleration and Simulation in Aircraft Technology Planning," *IEEE Transactions on Engineering Management*, Vol. EM-34, No. 4, pp. 244–251.

Batson, R. G. and Love, R. M. (1988), "Risk Assessment Approach to Transport Aircraft Technology Assessment," *AIAA Journal of Aircraft*, Vol. 25, No. 2, pp. 99–105.

Bell, T. E., ed. (1989), "Special Report: Managing Risk in Large Complex Systems," IEEE Spectrum, June, pp. 21–52.

Beroggi, G. E. G. and Wallace, W. A. (1994), "Operational Risk Management: A New Paradigm for Decision Making," *IEEE Transactions on Systems, Man, and Cybernetics*, Vol. 24, No. 10, pp. 1450–1457.

Black, R. and Wilder, J. (1979), "Fitting a Beta Distribution from Moments," Memorandum, Grumman, PDM-OP-79-115.

Book, S. A. and Young, P. H. (1992), "Applying Results of Technical-Risk Assessment to Generate a Statistical Distribution of Total System Cost," *Presented at the AIAA 1992 Aerospace Design Conference*, Irvine CA, February 3–6.

Chapman, C. B. (1979), "Large Engineering Project Risk Analysis," *IEEE Transactions on Engineering Management*, Vol. EM-26, No. 3, pp. 78–86.

Chiu, L. and Gear, T. E. (1979), "An Application and Case History of a Dynamic R&D Portfolio Selection Model," *IEEE Transactions on Engineering Management*, Vol. EM-26, No. 1, pp. 2–7.

Cullingford, M. C. (1984), "International Status of Application of Probabilistic Risk Assessment," Risk & Benefits of Energy Systems: Proceedings of an International Symposium, Vienna Austria, IAEA-SM-273/54, pp. 475–478.

Dean, E. B. (1993), "Correlation, Cost Risk, and Geometry," *Proceedings of the Fifteenth Annual Conference of the International Society of Parametric Analysts*, San Francisco, CA, June 1–4.

Degarmo, E. P., Sullivan, W. G., Bontadelli, J. A. and Wicks, E. M. (1997), *Engineering Economy*, Tenth Edition, Prentice Hall, Upper Saddle River, NJ.

Dienemann, P. F. (1966), *Estimating Cost Uncertainty Using Monte Carlo Techniques*, The Rand Corporation, Santa Monica, CA, January, RM-4854-PR.

Dodson, E. N. (1993), *Analytic Techniques for Risk Analysis of High-Technology Programs*, General Research Corporation, Santa Barbara, CA, RM-2590.

Fairbairn, R. (1990), "A Method for Simulating Partial Dependence in Obtaining Cost Probability Distributions," *Journal of Parametrics*, Vol. X, No. 3, pp. 17–44.

Garvey, P. R. and Taub, A. E. (1992), "A Joint Probability Model for Cost and Schedule Uncertainties," *Presented at the 26th Annual Department of Defense Cost Analysis Symposium*, September.

Greer, W. S. Jr. and Liao, S. S. (1986), "An Analysis of Risk and Return in the Defense Market: Its Impact on Weapon System Competition," *Management Science*, Vol. 32, No. 10, pp. 1259–1273.

Hazelrigg, G. A. Jr. and Huband, F. L. (1985), "RADSIM - A Methodology for Large-Scale R&D Program Assessment," *IEEE Transactions on Engineering Management*, Vol. EM-32, No. 3, pp. 106–115.

Henley, E. J. and Kumamoto, H. (1992), *Probabilistic Risk Assessment*, IEEE Press, Piscataway, NJ.

Hertz, D. B. (1979), "Risk Analysis in Capital Investment," *Harvard Business Review*, Vol. 9, pp. 169–181.

Honour, E. C. (1994), "Risk Management by Cost Impact," *Proceedings of the Fourth Annual International Symposium of the National Council of Systems Engineering*, Vol. 1, San Jose, CA, August 10–12, pp. 23–28.

Hutzler, W. P., Nelson, J. R., Pei, R. Y. and Francisco, C. M. (1985), "Nonnuclear Air-to-Surface Ordnance for the Future: An Approach to Propulsion Technology Risk Assessment," *Technological Forecasting and Social Change*, Vol. 27, pp. 197–227.

Kaplan, S. and Garrick, B. J. (1981), "On the Quantitative Definition of Risk," *Risk Analysis*, Vol. l, No. 1, p. 27.

Keeney, R. L. and von Winterfeldt, D. (1991), "Eliciting Probabilities from Experts in Complex Technical Problems," *IEEE Transactions on Engineering Management*, Vol. 38, No. 3, pp. 191–201.

Markowitz, H. M. (1959), *Portfolio Selection: Efficient Diversification of Investment*, Second Edition, John Wiley & Sons, New York, NY, reprinted in 1991 by Basil Blackwell, Cambridge, MA.

McKim, R. A. (1993), "Neural Networks and the Identification and Estimation of Risk," *Transactions of the 37th Annual Meeting of the American Association of Cost Engineers*, Dearborn, MI, July 11–14, P.5.1–P.5.10.

Ock, J. H. (1996), "Activity Duration Quantification under Uncertainty: Fuzzy Set Theory Application," *Cost Engineering*, Vol. 38, No. 1, pp. 26–30.

Quirk, J., Olson, M., Habib-Agahi, H. and Fox, G. (1989), "Uncertainty and Leontief Systems: An Application to the Selection of Space Station System Designs," *Management Science*, Vol. 35, No. 5, pp. 585–596.

Rowe, W. D. (1994), "Understanding Uncertainty," *Risk Analysis*, Vol. 14, No. 5, pp. 743–750.

Savvides, S. (1994), "Risk Analysis in Investment Appraisal," *Project Appraisal*, Vol. 9, No. 1, pp. 3–18.

Sholtis, J. A. Jr. (1993), "Promise Assessment: A Corollary to Risk Assessment for Characterizing Benefits," *Tenth Symposium on Space Nuclear Power and Propulsion*, American Institute of Physics Conference Proceedings 271, Part 1, Albuquerque, NM, pp. 423–427.

Shumskas, A. F. (1992), "Software Risk Mitigation," in Schulmeyer, G. G. and J. I. McManus (eds.), *Total Quality Management for Software*, Van Nostrand Reinhold, New York, NY.

Skjong, R. and Lerim, J. (1988), "Economic Risk of Offshore Field Development," *Transactions of the American Association of Cost Engineers*, American Association of Cost Engineers, New York, NY, pp. J.3.1–J.3.9.

Thomsett, R. (1992), "The Indiana Jones School of Risk Management," *American Programmer*, Vol. 5, No. 7, pp. 10–18.

Timson, F. S. (1968), *Measurement of Technical Performance in Weapon System Development Programs: A Subjective Probability Approach*, The Rand Corporation, Santa Monica, CA, Memorandum RM-5207-ARPA.

Wang, J. X. (1991), "Fault Tree Diagnosis Based on Shannon Entropy," *Reliability Engineering and System Safety*, Vol. 34, pp. 143–167.

Wang, J. X. (1996), "Complexity as a Measure of the Difficulty of System Diagnosis," *International Journal of General Systems*, Vol. 24, No. 3, pp. 257–269.

Wang, J. X. (2002), *What Every Engineer Should Know about Decision Making under Uncertainty*, CRC Press, Boca Raton, FL.

Wang, J. X. (2008), *What Every Engineer Should Know about Business Communication*, CRC Press, Boca Raton, FL.

Wang, J. X. (2010), *Lean Manufacturing Business Bottom-Line Based*, CRC Press, Boca Raton, FL.

Wang, J. X. (2015), *Cellular Manufacturing Mitigating Risk and Uncertainty*, CRC Press, Boca Raton, FL.

Wang, J. X. (2017), *Industrial Design Engineering: Inventive Problem Solving*, CRC Press, Boca Raton, FL.

Wang, J. X. (2019), "Complexity as a Measure of the Difficulty of System Diagnosis in Next Generation Aircraft Health Monitoring System," SAE Technical Paper 2019-01-1357, doi:10.4271/2019-01-1357.

Wang, J. X. (2019), "A Dynamic Fault Tree Approach for Time-Dependent Logical Modeling of Autonomous Flight Systems," SAE Technical Paper 2019-01-1358, doi:10.4271/2019-01-1358.

Williams, T. (1995), "A Classified Bibliography of Recent Research Relating to Project Risk Management," *European Journal of Operations Research*, Vol. 85, pp. 18–38.

Index

Printed in the United States
by Baker & Taylor Publisher Services

Printed in the United States
by Baker & Taylor Publisher Services